T0304958

hyper ⟶
efficient

Also by Mithu Storoni

Stress Proof

hyper → efficient

Optimize Your Brain to Transform the Way You Work

MITHU STORONI

yellow kite

First published in the United States in 2024 by Little, Brown Spark
An imprint of Little, Brown and Company
A division of Hachette Book Group, Inc.

First published in Great Britain in 2024 by Yellow Kite
An imprint of Hodder & Stoughton Limited
An Hachette UK company

1

Copyright © Mithu Storoni, M.D., Ph.D., 2024

The right of Mithu Storoni, M.D., Ph.D., to be identified as the
Author of the Work has been asserted by her in accordance
with the Copyright, Designs and Patents Act 1988.

All rights reserved. No part of this publication may be reproduced, stored
in a retrieval system, or transmitted, in any form or by any means without
the prior written permission of the publisher, nor be otherwise circulated
in any form of binding or cover other than that in which it is published and
without a similar condition being imposed on the subsequent purchaser.

A CIP catalogue record for this title is available from the British Library

Trade Paperback ISBN 9781399716024
ebook ISBN 9781399716031

Printed and bound in Great Britain by Clays Ltd, Elcograf S.p.A.

Hodder & Stoughton policy is to use papers that are natural, renewable
and recyclable products and made from wood grown in sustainable
forests. The logging and manufacturing processes are expected to
conform to the environmental regulations of the country of origin.

Yellow Kite
Hodder & Stoughton Limited
Carmelite House
50 Victoria Embankment
London EC4Y 0DZ

www.yellowkitebooks.co.uk

For Laurent

Praise for *Hyperefficient*

"Have you ever wished you could get in just one more hour of focused work, but your brain just wouldn't cooperate? *Hyperefficient* addresses one of the most vexing dilemmas knowledge workers face: the fact that the demand on our attention is limitless. This book offers strategies that make sustaining attention easier, helping you shut out distractions, and concentrate on high-level work."

—NIR EYAL, author of *Indistractable*

"A new approach to organizing your work—one that is designed for your human mind and body. Storoni teaches you how to find your flow and slow down so that you can get far. In a fast-moving world, *Hyperefficient* is a reminder that with self-awareness and a few environmental modifications, we can all be better at achieving our life goals. Such a fun and useful read!"

—AYELET FISHBACH, Professor of Behavioral Science and Marketing at the University of Chicago Booth School of Business and author of *Get It Done*

"Read this book! Mithu Storoni's unique strategy doesn't just preserve brain health and longevity, it promises to escalate mental performance to new heights and improve the way we work."

—DAN BUETTNER, National Geographic Fellow and #1 *New York Times* bestselling author of *The Blue Zones*

"One of the smartest 'smart thinking' books that I have read. Mithu Storoni's description of the brain's 'gears' provides an exceptionally elegant metaphor of the mind's workings, and powerfully explains how we respond to the pressures of the Information Age. It is beautifully written and packed with practical advice that should transform the ways we work—and rest. Whatever challenges you are facing, this book will help you to thrive."

—DAVID ROBSON, author of *The Expectation Effect*

"This is a very important and inspiring exploration and explanation of the mind, performance and our potential as human beings. This book holds the power to make a serious difference."

—JONNY WILKINSON CBE, Rugby World Cup Champion

"Whether caught in the eye of a thunderous storm or in the lethargic slip-stream of anxiety, Mithu Storoni has advice for you. Know where you want to arrive, settle on the right pace and connect your mind and body."

—DAME SANDRA DAWSON, Professor Emerita, University of Cambridge

CONTENTS

Contents

hyper ⟶
efficient

SECTION A

A New Kind of Efficiency

"What may emerge as the most important insight of the twenty-first century is that man was not designed to live at the speed of light."

—MARSHALL MCLUHAN[1]

Once upon a time, news traveled no faster than the pace of scurrying feet. Technology upended that world. It distorted time, contorted space, and gradually synchronized minds and brains across the globe into a fluid new reality.

On the 27th of August 1883, the residents of Milwaukee, Wisconsin, shuddered in terror when the mighty volcano Krakatoa in Java, Indonesia, emitted the loudest sound ever to be heard on Earth. The deafening blast was actually imperceptible to the people of Milwaukee, and yet they trembled as if Krakatoa had erupted right in their backyards.[2] A new communication technology called the telegraph was responsible for this peculiar dismembering of reality: it disseminated news of the disaster to sixty newspapers across the world, thereby shattering the immutability of space and time as the scaffold of existence. Human beings could now hop into a reality thousands of miles away, without needing hours or even weeks to arrive. Or as the philosopher Marshall McLuhan put it, "Today it is only too easy to have dinner in New York and indigestion in Paris."

Before the railroads arrived, Americans lived on "sun time." Each town in the United States had its very own brand of time, set by the position of the sun in the sky, and every degree change of longitude resulted in a time difference of four minutes. After the railroads arrived, Americans could no longer let the sun decide what time it was, as this created too much confusion. On April 19, 1883, *The New York Times* lamented how "fifty-six standards of time are now employed by the various railroads of the country in preparing their schedules of running times."[3] Human pace had now overtaken the sun's, and the sun would have to concede the reins of time to humans. By the end of 1883, the hills and valleys of time had flattened into a continuous American landscape, and "standard" time was born.

Once technology began to contort the coordinates of time and space, it played puppeteer to the trajectory of human existence. Railroad tracks defined *where* humans moved, and railroad timetables decided *when*. Humans consented to confine their movements—and, in turn, the patterns of their lives—to the departure and arrival times printed on a train schedule in exchange for a more efficient way of life. Soon the quest for efficiency would change the way humans moved within even tighter confines of time and space.

Over the course of the nineteenth century, scientists from Helmholtz to Huxley had shown that humans, like inefficient machines, waste energy. As the century neared its end, a sedulous Pennsylvania machine shop engineer named Frederick Winslow Taylor noticed that a worker's output could be substantially augmented if they did not stray beyond a strict choreographed sequence of movements while they worked. Taylor realized that the limiting factor in manual work was not *insufficient* physical effort but *wasted* physical effort. A precisely designed routine could accelerate an entire manufacturing process, and even make the work easier. Taylor was the first to study the science of work efficiency in an industrial workplace, rather than in a laboratory, and this made his findings immediately translatable to the real world. As his ideas spread across manufacturing plants in the United States, one plant, the Ford Motor Company, took the mission of improving efficiency to an entirely new level. Engineers at the plant experimented with a strategy that would turn the standard model of car manufacturing upside down. Instead of several engineers working on one car at a time as they had always done, one engineer worked on several cars in quick succession — by moving each on a conveyor belt in the form of an "assembly line". This erased the time wasted between jobs and significantly escalated output per unit of time. Their strategy proved so successful that the Ford Motor Company went from producing eleven cars in a month to churning one out every twenty-four seconds, and the assembly line took the stage as the global model for efficiency.

Little by little, all flavours of work were recast into an assembly-line template. The undulations of farm labor flattened into a continuous moving belt of mechanized agriculture, identical houses were assembled from a moving belt of parts, and even patients on hospital-clinic conveyor belts waited their turn to be "fixed."

In the 1950s, when a large portion of work moved from the hands to the head, many rural Americans flocked to urban America, seduced by the shiny, gilded image of white-collar knowledge workers working comfortably for better pay. But the factory model of work had hardly changed. Office workers sat in place and performed monotonous, almost automatic work on repeat; they clocked in every morning and out in the evening, fueled by caffeine and interrupted only by hunger's call to lunch. Even the office layouts were modeled after the factory floor, with desks crammed in rows like stations on an assembly line.

By 1970, the most dissatisfied workers in the country were no longer blue-collar workers assembling parts in a factory but white-collar clerical workers on desk jobs. Industrial age factories had metamorphosed into gleaming, shiny offices, and *desk workers' brains* were the new assembly lines. As a 1972 report by a special task force to the United States Secretary of Health, Education, and Welfare described it, "The office today, where work is segmented and authoritarian, is often a factory. . . . Computer keypunch operations and typing pools share much in common with the automobile assembly line."[4]

A linear, continuous, assembly-line configuration of

work emphasizes quantity over quality. When you flatten the natural fluctuation of your mental engine into a straight line, you prevent the lows, but you also prevent the highs. There is no room for flair or brilliance, or for out-of-the-box thinking. For well over a century, the majority of white-collar work did not demand complex thinking or exceptional creativity, so this was of no great significance. Human minds were mere cogs in the production machine; how *much* they worked mattered more than how *well* they worked. In an assembly line of moving minds, as in an assembly line of moving hands, quantity decided the bottom line.

It is a feature of the human condition that whenever we encounter a limit, we problem-solve our way around it by creating tools. When the pace of human hands hit a ceiling of maximum output, we supplanted human hands with automated machine parts. When the computational power of the human mind limited progress, we delegated its labor to calculators, then to computers, and then to super-computers. Machine intelligence can now do straightforward knowledge work more efficiently than humans. Automated trading has replaced human minds across Wall Street trading floors, legal firms are integrating chatbots into their legal service, and AI races ahead of eye surgeons to successfully screen patients for diabetic eye disease.

With AI gaining territory over lower-level-thinking jobs, the knowledge work spectrum is shifting toward an emphasis on idea generation, complex learning, and problem-solving. In his book *Exponential,* Azeem Azhar describes how a company's worth today depends more on its intangible assets than on its tangible ones. The algorithms in a search engine's platform are more valuable than the physical factories that manufacture its hardware, and the brand identity of a company like Apple holds more value than the physical cost of all its products. These intangible assets are realized through an exceptional level of "brainwork." Products have also grown so complex in the digital age that they are more difficult to conceptualize than to make. For example, a cell phone's manufacturing process is insignificant compared to the sophisticated thought process needed to design its software.

The old kind of factory-style continuous work creates the wrong kind of ecosystem in this new phase of the digital age. The assembly-line-flattened mind, without its peaks of brilliance and troughs of recovery, will not be optimally placed to flourish in a world where generating genius ideas and designing ingenious solutions define success. Efficiency is no longer defined by the quantity of output, but by its quality.

In the following chapters, I propose a new approach to efficiency for the age of AI-assisted knowledge work. I begin by searching for clues that tell us how our brains aspire to work when work is not imposed from the outside, and trace this trail deep into the brain's wiring to search for footprints of these inclinations. I then tie current neuroscientific thinking to these footprints to create a template for mental efficiency that taps into the brain's inherent makeup and hard wiring. As you journey through this book, you will discover how mental performance can soar to exceptional heights if, instead of imposing the rhythm of assembly-line work on your brain, you impose the rhythm of your brain on your work.

1

POWER LAWS

"The very remote future [infinity] of . . . the aeon prior to ours . . . was our Big Bang."

— SIR ROGER PENROSE[1]

When mathematician and Oxford professor Sir Roger Penrose accepted his Nobel Prize in 2020, he spoke of "a crazy theory" of his: that before the Big Bang, there had been another universe just like ours that had burst forth from another big bang. According to his theory, there was an aeon before ours that was just like ours, and an aeon before that, and another before that, ad infinitum.[2]

Penrose's theory would resonate with many communities around the world, including the Ju'/hoansi of Botswana, who think of time as a rhythmic cycle, with neither a beginning nor an end. Life happens in a plexus of multiple repeating rhythms "characterized by the predictability of the seasons and the metronomic periodicity of the movements of the sun, stars, and moon," anthropologist James Suzman writes in *Affluence Without Abundance*.[3] In rhythmic time, the overall pace of life is set by nature's cycles; it cannot be accelerated with human tools. The sun's rise and fall, for example, defines the cycle of day and night, and try as you might, you cannot make the sun rise or fall sooner or later than it inevitably will.

Within each cycle, time passes at a variable rate, expanding when there is little to be done, and shrinking when activity soars. You get a sense of what this feels like if you have no visible indicator of time passing and your thoughts and perception of the world determine its apparent pace. Some periods seem to pass quickly, and others more slowly. You cannot easily slice these expansions and contractions of time into neat segments of steady assembly-line progress. In a world that spins on rhythmic time, time does not shape how you work; rather, your work shapes how you perceive time.

Curiously, this perceived nonuniform time may set the pace for our minds and bodies more than we think. In an extraordinary experiment, researchers at Harvard University measured how quickly thirty-three healthy volunteers recovered from superficial skin wounds sustained from cupping therapy in three different environments of *perceived* time. Each volunteer sat in front of a clock that had been manipulated to run half as fast as real time, twice as fast as real time, or at the pace of real time. The results were astonishing. When the volunteers believed time was passing faster than it actually was, *their wounds healed faster.*[4]

Stone Age Economics, written by the late anthropologist Marshall Sahlins, is a rich anthology of firsthand accounts of hunter-gatherer community life from different corners of the world.[5] The accounts paint a rhythmic pattern of work and rest such that people would work hard for part of the day and then spend a longer time in relative leisure. If a day was spent hunting, it would be followed by a couple of days of repose. Once everyone completed their work, they were not compelled to fill their ample free time doing any more.

Despite the persistent uncertainty and challenges of hunter-gatherer life, there was surprizingly little inclination to spend many hours embroiled in backbreaking work, and intense work dominated no more than short bursts of time. As Czech anthropologist Leopold Pospisil, who did extensive research on the Kapauku people of Papua, describes in Sahlins's anthology, "Only every other day is supposed to be a working day. Such a day is followed by a day of rest in order to 'regain the lost power and health.'" When their work called for hard labor for longer periods, "they [would] relax for a period of several days, thus compensating for their 'missed' days of rest."[6]

In another account, ethnologist Martin Gusinde describes the hunter-gatherer Yámana people of Tierra del Fuego,

South America, in the 1920s: "The Yámana are not capable of continuous, daily hard labor, much to the chagrin of European farmers and employers for whom they often work. Their work is more a matter of fits and starts, and... they can develop considerable energy for a certain time. After that, they show a desire for an incalculably long rest period during which they lie about doing nothing."

Almost every account in Sahlins's anthology tells of a rhythmic pattern of working, in which an intense burst of work is followed by periods of lighter work and rest. Such a pattern could be imagined as a kind of *power law*: a mathematical relationship between two things in which a change in one gives rise to a change in the other according to some power. In this case, as the laboriousness of the work rises, the time spent doing it falls, so the bulk of the time is spent doing moderate to light work. If we plot it as a graph, the power law relationship would look something like this:

POWER LAW CURVE

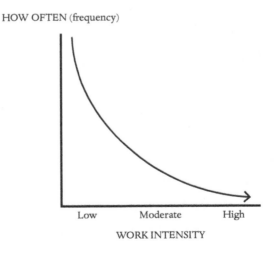

HOW OFTEN (frequency)

| Low | Moderate | High |

WORK INTENSITY

Power Laws

This way of working seems to be typical of diverse hunter-gatherer communities, irrespective of geography and independent of external pressures of weather, terrain, or climate. As Sahlins's work demonstrates, and as anthropologist James Suzman also points out in *Affluence Without Abundance*, hunter-gatherer communities — even those that still survive today — seem to have an innate propensity to work in this way.

In 2013, an American research group fitted forty-four traditional hunter-gatherers from the Hadza community in northern Tanzania with Global Positioning System (GPS) units as they went about foraging for food in their usual way. Almost half of their foraging bouts followed what is known as a Lévy walk: a power law pattern that roughly translates to walking short distances most often and long distances far less frequently. Perhaps even more intriguing, this pattern of walking was seemingly independent of the features of their environment. In other words, the Hadza hunter-gatherers did not walk in a power law pattern because they needed to; they did so because they chose to.[7] The Hadza are not unique. Other hunter-gatherer societies such as the Me'Phaa of Mexico and the Cariri of Brazil also demonstrate Lévy walk patterns.[8]

The instinct to hunt and forage in a power law pattern may have conferred an evolutionary advantage on our ancestors by making them less susceptible to life-threatening exhaustion and hunger when stranded in remote and alien landscapes.[9] If you have to locate a source of food in the wild with no prior knowledge of the terrain, the most efficient way to do so is by following a power law pattern: if you look for local sources first and venture out on long and exhausting expeditions only when absolutely necessary, you preserve energy for longer. The agricultural revolution and then the industrial revolution ironed over the instinct to behave and work in this way.[10] The uniformity and order necessary for successful agriculture erased all spontaneity and fluctuations. Assembly lines simply continued and amplified this trend.

The blueprint of the instinct to work and behave in bursts, in a power law pattern, however, intractably endures and is even woven into the tapestry of our inner experience. Searching for a memory is a form of foraging for the mind, and the same pattern of foraging seen in the Hadza, the Me'Phaa, and the Cariri emerges in the brain when we scour its shelves for a memory that has long been filed away.[11]

One compelling fragment of evidence for power laws embedded in the brain's dynamics is seen in newborn mammals. Being awake is labor-intensive for newborn brains, which are encoding new information and galvanizing neural connections at dizzying speeds. Sleep provides respite. Research has found that newborn rats flip randomly between sleep and awake states until they are around two

weeks old, when the periods they spend awake start to follow a power law.[12] They are awake for short bouts most of
the time, and for long bouts less often — a pattern that will
be familiar to parents of infants, who often nap in one-to-two-
hour intervals throughout the day.

Even adult humans living in a digitalized and predictable urban world seem to want to move and rest in a power
law pattern when left to their own devices. Actigraphy
studies have shown how we spontaneously pause to rest in
between bursts of physical activity in a power law pattern, taking short breaks frequently and long breaks less
often.

If power laws persist across climates and people and
lands, if they choreograph how we explore the world
around us and the depths of our own minds, if they carve
our innate patterns of work and rest, could they also hold
the secret to working better with our brains?

In 2006, researchers[13] from Hungary, Portugal, and the
United States raked through the archives of the Darwin
Correspondence Project, the Freud Museum of London,
and the Einstein Papers Project to see if there was a pattern
in the way Darwin, Freud, and Einstein — three of the most
celebrated minds of the twentieth century — answered
their letters. In those days, letters were a form of knowledge work. Many scientists did not work in "laboratories"
dedicated to spawning groundbreaking theories; rather,

they did such thinking at home, on a writing desk. Profound scientific debates and even a form of peer review took place using the medium of letter writing. Darwin's frequent letters to friends and colleagues chauffeured his mental voyage toward the theory of evolution.

Had Darwin, Freud, and Einstein replied to each letter as it arrived, there would be no obvious pattern to their letter writing. But this was not the case; the researchers found that there *was* a pattern. The time interval between receiving a letter and replying to it was often short and only occasionally long. Instead of writing like workers on an assembly line, with letters moving at constant speed on a conveyor belt, Darwin, Freud, and Einstein worked in rhythmic bursts with power law undertones.

Power laws are abundant in almost every facet of our natural environment, from the weather to ocean waves, from craters on the moon to volcanic eruptions, earthquakes, and avalanches. We also inadvertently incorporate power laws into human creations: cities and towns show power laws, as do corporations, features of the internet, and language. The power law patterns of the world very likely imprinted themselves upon us during our evolution and inspired power law patterns in our own neural and behavioural dynamics.

TECHNOLOGY AND TIME

When we use technology to improve a process, we target the slowest links in the chain because they are what hold the process back. We make the slow bits faster to catch up with the fast bits, so instead of having slow and fast phases, the entire process happens at a constant pace. This was the philosophy behind Ford's assembly lines. Since the pace of events sculpts our experience of time, time now passes at a constant speed for anyone observing the process.

Technology changes the coordinates of time so that a new kind of time emerges: *technological time.* This time is linear, regular, and set by the pace of technological chains. In most knowledge workspaces, workflows bend to meet this superhuman speed of information transfer. The whole office tries to synchronize itself to this pace, and technological time even follows workers all the way into their homes, burrowing into Zoom calls, text messages, and emails.

You can sense the magnitude of this effect if you find yourself in a world that technology has yet to fully conquer. A couple of years ago, I traveled to the remote mountains of Sardinia to meet explorer and *National Geographic* Fellow Dan Buettner to work on the Netflix series about his work, *Live to 100: Secrets of the Blue Zones.* We filmed in Seulo, a living museum of centenarians, where twenty-nine of its approximately eight hundred inhabitants were over a hundred years old.

Hard work is integral to life in Seulo, a town with a strong pastoral culture. Goatherds, even the centenarians among them, rise at the break of dawn and take their goats up to the mountains, no matter the season. Later, when the sun has risen high, they make their way back down to feast on a homemade meal. The afternoon is spent in gentle, satisfied existence, sharing news and gossip with neighbors and friends as they sit outside in the fresh, open air.

Many homes did not have a television or broadband internet, and the mobile phone signal was precarious. There was also significant day-to-day uncertainty. The terrain was hilly with few boundaries, the goats could wander anywhere, and every moment could bring something unexpected. Food was grown and gathered at nature's mercy, the weather decided how you would spend your day, and you had to change your plans if there was rain. Problem-solving was an inescapable feature of daily life. Yet despite the precarity inherent in their lifestyle, I saw few signs of stress or worry.

On my visit there, I could not help noticing that daily life passed by in bursts of intensity in a kind of power law pattern reminiscent of our hunter-gatherer ancestors.

Moments of intense mental and physical work existed, but they seemed to always be followed by a nap or a long interval of time spent gazing gently at the world passing by. You might have to chase a goat up a hill one moment, and spend the next hour lazily watching over your herd. The absence of technology meant that *humans* defined the speed of time. As they moved about, time followed in their footsteps. Once work was over, time slowed down too.

The people of Seulo score better in many health metrics than their counterparts in other parts of Sardinia and in Italy as a whole, with both mental and physical performance remaining optimal well into old age. Could their pattern of work and life be contributing to their extraordinary mental fitness and longevity?

Intense mental work depletes resources and generates mental fatigue much faster than light mental work. A "power law" mode of working—where you work intensely for short periods, and your work is punctuated with breaks and followed by rest and recuperation—protects you from chronic mental strain and its damaging consequences. But while working in rising and falling tides comes easily to the people of Seulo, it does not in itself guarantee prodigious mental output in a knowledge work environment. For rhythmic mental work to produce exceptional output, every interval of hard work would need to be optimized so it is consistently of the highest quality. For that to happen, we need one more ingredient.

THE POWER OF EFFICIENCY

In 2019, Eliud Kipchoge, a thirty-four-year-old long-distance runner from Kenya, broke every marathon record in human history when he completed a 26.2-mile course in less than two hours. But his astonishing feat never made it into the record books because he had used a special tool to help him. He didn't run with a jet propulsion backpack or use roller skates, nor did he take performance-enhancing drugs or wear smart sneakers. His tool was ingenious in its simplicity.

On the day of his race, a silver electric car with a formidable-looking apparatus strapped to its roof drove ahead of Kipchoge and shone a bright green laser beam on the ground in front of him while he ran. It acted as a pacemaker, as a metronome that orchestrated Kipchoge's every footfall, and it kept him at a steady pace of around two minutes and fifty seconds per kilometre, the most efficient running speed for achieving his goal. A faster pace increased the risk of exhaustion and a slower pace of losing the challenge. Kipchoge finished twenty seconds under the two-hour mark, and his triumph was a reminder of how a human body's performance can reach colossal heights by artfully managing its pace.

In many ways, the track was to Kipchoge what information is to our brains, and mental efficiency, like running efficiency, is a balancing act. You want your mental engine to work fast enough to make good progress but not so fast that you suffer wear-and-tear damage. Much like Kipchoge's optimal pace, there is an optimal "sweet spot" of mental pace where efficiency peaks.

But this poses a conundrum. By its definition, a power-law-like way of working is inherently rhythmic and entirely antithetical to working at a consistent, albeit efficient, pace. Also, *quality* and *efficiency,* though not mutually exclusive, do not always coexist. You may be an exceptionally efficient coder producing mediocre software, or you may arrive at a genius product idea after twenty years spent getting nowhere. If a rhythmic way of working produces excellence and a pacemaker maximizes efficiency, is there a way to work that will not sacrifice one for the other, but instead combine the two to forge an almost *hyperefficient* mode of working?

A setup that marries a power-law-like way of working with Kipchoge's pacemaker would need to do two things. First, the pacemaker would adjust your pace to suit the specific kind of mental work you are doing. Second, the pacemaker would change your pace in a power law pattern, keeping your pace high for only short bursts of time and low for much longer. This would preserve rhythmic peaks of brilliance while sharpening those peaks to be as efficient as possible. In the next section, I will describe how your brain is wired with an ingenious circuit that does exactly this.

2

YOUR GEAR NETWORK

"My mind is like a racing engine, tearing itself to pieces because it is not connected up with the work for which it was built."

— SIR ARTHUR CONAN DOYLE[1]

Nestled at the root of your brain, on either side of a cavity filled with spinal fluid, sits a small cluster of brain cells. The cells are full of the neurotransmitter *norepinephrine*, which makes them look blue. Hence they have been given the name *locus coeruleus*, the Latin term for a blue dot.[2] Your blue dot is your brain's hub of norepinephrine.

The brain cells in your blue dot fire "bullets" of norepinephrine that scatter across the brain's vast landscape through an intricate network, the "blue dot network." These bullets have an intriguing effect: they change the brain's entire configuration depending on the pattern in which they are fired.

Each configuration endows your brain with unique abilities and sets the pace at which it works. When your blue dot network sends out a fast stream of norepinephrine bullets, your thoughts speed up. When it fires more slowly, your thoughts slow down.

The blue dot is usually referred to by the Latin term *locus coeruleus*, which is often shortened to **LC**. The LC, together with its nerve trails that carry and dispatch norepinephrine, is referred to as the LC-NE network in the scientific literature. The LC-NE network doesn't work in isolation: it has a close relationship with both dopamine-based and acetylcholine-based systems. For the purposes of this book, I will refer to the whole LC-NE network and its second-order effects with dopamine and acetylcholine as the *blue dot network*. (The three gears I subsequently describe refer loosely to the states associated with three levels of tonic and phasic LC activity, with gear 2 corresponding to the highest levels of phasic LC activity.)

In this way, your blue dot network expertly controls your brain's configuration, changing your brain's state simply by changing the way it fires.

Your Gear Network

In the context of knowledge work, it helps to imagine your blue dot network as a gear system that puts your brain into three main modes: *slow, medium,* and *fast.* You shift gears to change mental pace. **Gear 1** puts you in slow mode, **gear 2** in medium mode, and **gear 3** in fast mode.

Gear 1 is like sitting in an armchair by the fire: it helps you rest, recover, and daydream after hard labor. Gear 2 lets you work comfortably and efficiently. It calibrates your brain to optimally focus, learn, solve, and analyse. The gear 3 state of mind is like sprinting: it gives you a jolt of energy, but it tires you out if you stay in it for too long.

Emerging research shows that the brain is put under strain if it spends long periods in gear 3. At the same time, a gear 1/gear 2 state may offset the harm caused by being in gear 3. In 2021, a Canadian research team demonstrated how mice engineered to be more susceptible to brain disease were less likely to develop disease if their blue dot network was made to fire in a gear 2 pattern.[3] Gear 3 enhanced wear-and-tear damage, but gear 2 seemed to "rescue" the brain cells from some of this damage. If confirmed in humans, this finding has an astonishing implication. It suggests that the fast-paced gear 3 mental state need not be entirely avoided — it can be embraced in short doses, so long as this is counterbalanced by enough time spent in a lower gear. This adds further support to the notion that a power-law-like pattern, with only short times in gear 3 and longer times in gears 1 and 2, would seem the most logical approach to mental work.

Intriguingly, your blue dot network may itself tend toward such a pattern: earlier, I described how newborn rats begin to show a power law pattern in the way they behave at around two weeks. The blue dot network seems to play a central role in inducing this change.[4,5] In other words, the blue dot network comes very close to fulfilling the criteria I outlined earlier: it can set the pace of your brain to suit the specific kind of mental work you are doing, *and* it has the capacity to change your pace in a power-law-like pattern.

At a superficial level, your blue dot network changes the *pace* at which your mind runs. Digging deeper, each gear changes the precision of your focus, the volume of data you process, and the overall efficiency of your mental engine. This artful control of mental pace bends Darwin's quality and Kipchoge's efficiency into a kind of *hyperefficiency.*

GEAR 1

Best for: Recharging your mental engine, resting, daydreaming

Think of how you feel when you first wake up in the morning, or when you are sitting lazily on a park bench late in the afternoon, watching the world drift by. Your attention gently grazes the surface of everything around you, so you might notice a leaf here, or a passing thought there, but you move on.

This state of mind—when you feel relaxed and no single thing holds your attention—is what it feels like when your brain is in gear 1. Gear 1 is a "slow power" mental state: a state of mind that is not revved enough to go through complex information. Your attention is not powerful enough to stick with anything for long.

Using a photography analogy, you have a panoramic view, but the image is a little fuzzy and you cannot zoom in to anything.

This panoramic perspective makes it difficult to attach your focus to a single task, and this is why the gear 1 state of mind is indispensable to mental work: it lets you unplug your attention and rest your mind. It also helps you wipe your mental slate clean so you resume your task with a fresh approach and can see it from a new angle.

With a detached grip on the outside world, your mind becomes more receptive to thoughts *inside* your head. You don't fix your attention there either, but when idea bubbles rise into your conscious mind from the cauldron of your subconscious, you are more likely to acknowledge them. It creates the right environment for *a-ha!* moments of inspiration.

GEAR 2

Best for: Mental work, concentration, learning, problem-solving, critical thinking, and creativity

The unique quality of gear 2 is that it bestows you with what is almost a magic power in the world of mental work: *the ability to focus attention.*

This happens because your prefrontal cortex, the region of your brain that sits behind your forehead—and is the master seat for all mental work—is fully engaged in gear 2.

Your prefrontal cortex, the part of your brain behind your forehead, is the master seat of all mental work—it enables you to think, imagine, predict, analyze, decide, solve problems, and, most important of all, focus attention. Your prefrontal cortex works best at medium levels of norepinephrine, exactly at the sweet spot of gear 2, when phasic LC activity is highest.

Using the analogy of a camera lens, in a gear 2 state of mind, you bring the target of your attention into sharp focus, whilst the background is blurred. Your beam of focus can face outward *or inward,* and it can also narrow and widen without you losing focus.

Consider how your focus changes across three activities that are performed optimally in gear 2: reading a book for pleasure, analyzing a report, or brainstorming ideas. When you're buried in an escapist novel, your mind's pace feels slower, and you switch between focusing on the words on your page one moment and letting your attention wander the next. If you have to read a heavy, formal report, your mind's pace feels faster, your attention narrows, and your focus fixates on the report without wandering. When you brainstorm ideas for a product you are excited about, your mind's pace feels faster still. You alternate between widening your attention to capture many ideas and narrowing your attention to zoom in to each idea.

Your attention is narrow and intensely sticky at the core of gear 2. As you edge away from its core and drift into its peripheries, your attention — while still good enough to focus — *widens* and becomes *less sticky.* Your mind's pace subtly changes, too: as you approach the boundary with gear 1, your mind feels slightly *less energetic,* and as you near the boundary with gear 3, it feels slightly *more energetic.* This is a metaphorical way of looking at what happens but helps you visualize the dynamics.

A *low-energy* gear 2 state

Best for: Spontaneous creativity

When you're in gear 2, near the boundary with gear 1, you are in what I will describe as a *low-energy gear 2* state. Your attention is slightly less sticky. You can actively detach your attention now and again from what you're working on and let it gently drift. When you detach your attention, it gives you the opportunity to wipe your mental slate clean, just like in gear 1, and to gain perspective on your task. As your attention drifts, it ventures inward, into your mind. Once there, it casually saunters around fragments of loose threads and ideas, inadvertently stumbling upon spontaneous insights: gems it would have missed had it moved at a faster pace with an unrelenting, narrow beam of attention. In a low-energy gear 2 state, you can let your attention wander one moment and quickly narrow it the next to bring one of these emerging insights into sharp focus. This part-detached, part-focused state is ideal for spontaneous creativity; you forage for ideas and can filter what you foraged with focus.[6] Its main difference from gear 1 is that your mental engine has the power to focus, so you are able to shift between focusing and defocusing easily, viewing the world through a narrow and a wide-angle lens in quick succession.

A *high-energy* gear 2 state

Best for: Learning complex concepts, divergent thinking, brainstorming

When you're in gear 2, near the boundary with gear 3, you are in what I will describe as a *high-energy gear 2* state. Here norepinephrine levels are at their highest levels within gear 2, which is why you feel energized. Norepinephrine has an intriguing property: it makes soft signals louder.[7] Peripheral details you could not see before now show up on your radar, boosting your ability to think laterally, see new perspectives, brainstorm ideas, and make connections as you learn. This happens both in the outside world and inside your mind.

This energized state of focus creates optimal conditions for learning: the raised norepinephrine levels in this state enhance learning through routes that include increasing energy supply to brain cells and encouraging new connections among them. This high-energy gear 2 state is in turn *created* when you find yourself driven by the fire of intrinsic motivation (more on this in chapter 7) and passionate curiosity. This feature makes this mental state one of the most productive and fulfilling states to be in while you work.

It has long been observed that being in a gear 2 state of mind feels good for most people. It is possible this is because when you are in it, your mind assumes the most compatible configuration for you to be able to unlock, discover, and comprehend (through learning) the world around you and become a part of that world. The author James Hilton captured this idea in his book *Lost Horizon*, in which people live to be over two hundred years old in a remote location, Shangri-La, by aspiring toward this "moderate" state of mind.

GEAR 3

Best for: Responding quickly in critical moments

In gear 3, your blue dot network is firing norepinephrine bullets in quick succession and your mental engine drills through information at a rapid pace. Your prefrontal cortex is partly offline. As a result, your refined mental abilities are heavily compromised, but your "crude" mental abilities—ones that don't require any thought—are enhanced. Your typing speed will skyrocket, but you won't be able to thoughtfully analyse what you type. Using the metaphor of the camera, your image looks like a blur of movement with fuzzy lines and no details.

In this state, you can execute anything that feels automatic—or that you have rehearsed so well you don't need to think about it—extremely fast without wasting time on thinking. This makes gear 3 ideal for responding to emergency, time-critical situations, when you don't want to be distracted by emotions or analysis. Medical staff attending a cardiac arrest perform better in gear 3 than in gear 2, as do athletes at critical moments of play.

This gives the gear 3 mode a critical role in AI-driven workspaces; in a world where a glitch in an automated digital system can have multiplicative and massive effects on a global scale within moments, acting quickly can prevent a disaster of immense proportion.

As your blue dot network increases its firing and you go deeper into gear 3, more data gushes into your mind and your mental engine struggles to separate signal from noise. The outside world feels overwhelming and your inner world floods with racing thoughts. The torrent of norepinephrine released by your blue dot network gives you energy and numbs pain and fear, but it also hampers your judgment. In a knowledge work setting, these attributes place you at a disadvantage: you are more likely to misunderstand, use bias, jump to conclusions, miss nuance, and make hasty decisions.

Your brain's emotion-processing pathways have a direct conduit into your blue dot network. Emotional triggers easily raise your gear, and they amplify and prolong the effects of anything else that raises your gear. Emotional arousal can topple you into gear 3 and keep you there, preventing you from downshifting at the end of the day. This is why falling asleep can feel difficult after experiencing intense negative emotions over the course of the day, even if the reason for the emotions no longer exists.[8]

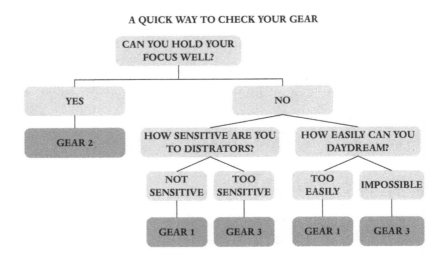

A QUICK WAY TO CHECK YOUR GEAR

A QUICK VIEW

Your brain is much like the surface of an ocean: it never stays still. It is always dynamic, its state fluctuating with shifts in the wind and the currents. Just as an ocean's surface can become calmer or more turbulent as its wave patterns fluctuate, your brain, too, is constantly flickering across all three gear states. Think of how a focused mind (gear 2) needs to detach (gear 1) every now and again to wipe its slate clean, or how a racing, distracted mind (gear 3) can be jolted into focus (gear 2) when some piece of salient information grabs your attention. This is why, throughout the book, when I describe being in a particular gear, I refer to being **predominantly** in that gear state.

Here is a short summary of the metaphorical gear model:

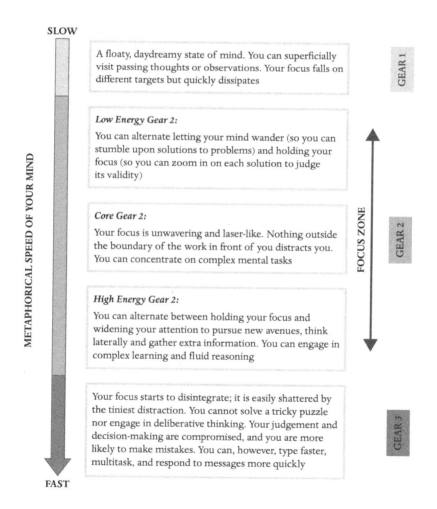

SLOW

A floaty, daydreamy state of mind. You can superficially visit passing thoughts or observations. Your focus falls on different targets but quickly dissipates

GEAR 1

Low Energy Gear 2:

You can alternate letting your mind wander (so you can stumble upon solutions to problems) and holding your focus (so you can zoom in on each solution to judge its validity)

Core Gear 2:

Your focus is unwavering and laser-like. Nothing outside the boundary of the work in front of you distracts you. You can concentrate on complex mental tasks

FOCUS ZONE

GEAR 2

High Energy Gear 2:

You can alternate between holding your focus and widening your attention to pursue new avenues, think laterally and gather extra information. You can engage in complex learning and fluid reasoning

Your focus starts to disintegrate; it is easily shattered by the tiniest distraction. You cannot solve a tricky puzzle nor engage in deliberative thinking. Your judgement and decision-making are compromised, and you are more likely to make mistakes. You can, however, type faster, multitask, and respond to messages more quickly

GEAR 3

FAST

METAPHORICAL SPEED OF YOUR MIND

3

WHAT IS YOUR GEAR PERSONALITY?

"When the going gets tough, the tough get going."
—JOSEPH P. KENNEDY SR.

How easily you slide from one gear to another depends to some degree on your *gear personality*. Some of us have *springy* gears that shoot up at the gentlest nudge, while others with *stiff* gears need a bigger shove. Most of us have a gear personality that lies somewhere in between. Gear personality differences help to explain why everyone tolerates different thresholds of uncertainty, why some people have more difficulty getting "in the zone" than others, and why the same environment can make one person jittery and another person yawn.

The variation in gear personality is on full display on the trading floors of large investment banks. On one side sit battle-hardened traders gazing calmly at endless strings of red and green-colored numbers as they flash across enormous screens. On the other side sit cautious mathematicians contemplating the possibility of a tiny error in the coding algorithm or risk calculation, even if it is several decimal points in. The traders and the mathematicians may both be in gear 2, but each uses a different level of stimulation to get there.

To effectively control your gears, it helps to know where you stand on the gear-personality spectrum. Do you like to be in environments with more (or less) stimulation? Do you work better under pressure? Are you more stressed by trivial matters than others seem to be?

We are innately drawn to ecosystems that match our gear personality unless this is overridden by other factors. Many doctors enter medicine with noble intentions and many entrepreneurs set out to launch world-changing products, only to discover an unbearable mismatch between the job and their gear personality along the way. Understanding your gear personality can help you identify the kinds of environments where you are most likely to thrive.

A "STIFF" GEAR PERSONALITY

If you like pressure, need deadlines to motivate you, and feel energized rather than drained by challenges, your gear is likely to be at the *stiff* end of normal. You need more of a push to reach gear 2 and to stop yourself from sliding back down to gear 1 once you are there.[1]

My friend PNC, a successful trader at one of the world's most successful Wall Street investment banks, offers a good example of a "stiff gear" mindset. He trades in millions of dollars and spends his days making high-stakes decisions, but he is so unfazed by the pressure at work that he feels lethargic enough to catch a quick nap after lunch every now and again. I remember the time PNC told me he wanted to take up yoga. For most people, learning the poses is a yearslong process. PNC challenged himself to learn to stand on his head without support in just a few days. PNC explained to me that he doesn't get pleasure from danger but from *exerting control* to avert it. Since the threshold at which uncertainty unnerves him is higher than average, PNC rarely feels the impetus to exert control in ordinary circumstances, and he needs highly uncertain environments, such as on a high-intensity trading floor or while skiing down Mount Fuji under dubious safety conditions (which he often does), to thrive. These activities would feel deeply uncomfortable for most people, but for PNC they provide just the right level of challenge that he can overcome by exerting control.

A special case: ADHD

Several research groups have suggested that some adults with attention deficit disorder tend to be lodged in gear 1 and find it hard to elevate themselves into gear 2. According to this framework, "hyperactivity," "sensation-seeking," and "risk-taking" are manifestations of their attempts to raise their gear; they need more stimulation and more uncertainty to nudge their gear upward.[2] Once in gear 2, their focus is optimal—often better than that of their non-ADHD peers.

Hunter-gatherer modeling experiments have shown that the presence of ADHD within a population may have helped them survive through change and challenge. People with ADHD were the risk-takers: they explored the unknown and thirstily sought knowledge. They did not waste time analysing decisions at critical moments. If *everyone* in a population had these traits, the whole population would be exposed to risk all the time. But if only a minority had them, these individuals could safeguard the entire population without exposing everyone to risk.[3] Although ADHD may be disadvantageous in the man-made, predictable world of knowledge work we inhabit, it may confer robustness on the human species, maximizing our survival should disaster ever strike on a massive scale.

Many high achievers with ADHD learn to identify and avoid the kinds of activity that leave them bored and distracted and instead find what interests and inspires them enough to excel. Rugby champion James Haskell, for instance, has referred to his ADHD as a "superpower" that guided him toward rugby, and he credits his ADHD for his prodigious athletic achievements.[4] Some other famous names with presumed or reported ADHD include Leonardo da Vinci, Richard Branson, Jamie Oliver, and Michael Phelps.[5]

A "SPRINGY" GEAR PERSONALITY

If you feel your gear leap up with the slightest nudge and you struggle to bring it back down, your gear is likely to be *springy*. Springy gears shoot upward with the lightest stimulation, often overshooting past gear 2 into gear 3. You react to a whisper as you would to a roar and to a text message beep as you would to a fire alarm—and once your gear shifts up, it is difficult to dial back down. Lower levels of stimulation and quiet, slower environments feel more comfortable for those with springy gears.

Your gear personality can be influenced by past experiences and hence can change with experience. A soldier might need a higher level of ambient stimulation to reach the right mental space for a while after returning from combat. Someone from a small town may initially feel overwhelmed by the noise and bustle of a big city before adjusting. As the British-Canadian psychologist Daniel E. Berlyne wrote in 1960, "The quantities of excitement... and environmental vicissitude that can be tolerated by the New York taxi driver and the Balinese farmer... the town mouse and the country mouse... will clearly differ."

Environment-induced change in gear personality can sometimes be insidious. As we grow accustomed to the rising levels of attention-grabbing stimuli in our environment, we start to depend on them to raise our gear and lose the skill to do this by ourselves. It becomes harder to anchor your attention to normal, everyday things if you always have your attention anchored for you. This may explain why some research has tied screen time at the age of eighteen to the diagnosis of ADHD by the age of twenty-two.[6,7]

Luckily, the problem holds the solution. If your gear personality can change in one direction, it can also change in the other. You can become better at regulating your gear and raising yourself into gear 2 simply by *practicing* paying attention.[8,9] Focused-attention meditation has long been used for training attentional control, and in the digital age there are digital options for these practices. In one trial, using a focused-attention meditation app improved subjects' ability to pay attention (and hence enter gear 2) in just six weeks.[10]

Once you have identified your particular gear personality, it helps to consider a second factor: your personal accelerators. What especially rattles you? What do you fear? What pushes your buttons more than anything else? Your answers to these questions will identify the triggers unique to you that excessively raise your gear.

PERSONAL ACCELERATORS

When BK was a child, he struggled to pay attention at school. He would disturb other students, refuse to learn, play games at his desk, and frustrate the teacher. When his teacher complained to his parents, he changed strategy. He would pretend to pay attention and sit very still while his mind sailed on magnificent adventures crafted by his imagination. His grades continued to plummet. When his father casually handed him a textbook targeted at older children one evening, he unwittingly changed the trajectory of his son's life. Suddenly, BK, who couldn't ever sit still, sat captivated by the contents of the book, his focus unwavering for over an hour. BK's teachers subsequently calibrated their teaching to this level of challenge, and BK graduated high school as one of the highest achievers in his school's history. BK seemed to need a significant challenge to raise his gear, but once he arrived in gear 2, his mental performance was exceptional. This pattern could also be traced in other areas of BK's life. He felt no fear when he once had to face down an armed burglar, and he was exhilarated when swimming with sharks for the first time. BK's story so far would suggest that he falls into the stiff gear category, but there is a catch. BK went from being the calmest person in the room to the most anxious if he ever had to catch a train or meet a deadline. BK's stiff gears became *hyper-*springy in the face of one personal accelerator: working to a time limit.

Today BK works at an IT company and excels at his job. Early in his career, BK realized how tasks that increased his sense of control, like organizing and ordering, acted as an antidote to the anxiety of meeting deadlines, and he threw himself into these activities even when they added to his workload. As he rose in seniority, he had fewer deadlines, and he thrived at overcoming difficult challenges. He now enjoys the reputation of being one of the strongest leaders in his firm.

We each carry a unique bundle of wiring, part donated by nature and part carved by nurture, which colors the way we respond to the world. A person who happily skis down black runs might be terrified of riding a horse because they fell off one as a child. Someone who is easily distracted by persistent thoughts and ruminations in their own mind might be immune to the distractions of a noisy office.

We would have long disappeared as a species had we been born with identical gear personalities. Human prefer-ence for *both* high- and low-uncertainty occupations has enabled us to master a range of challenges, which contrib-uted to survival and the advancement of civilization. Unflappable calm when facing mortal danger would have led to victory over predators in the savannahs of Africa, and satisfaction with a slower pace of life fueled the desire to pursue an unremarkable shrub. More recently, the need for adventure and excitement spurred expeditions to the unknown, and a preference for quiet contemplation catalysed the conjuring of mathematical laws and theorems.

It is possible that as the world continues to change, people with different gear personalities will adapt in different ways. Some will thrive in a world of acceleration, uncertainty, and excessive information, but for some others, such a world could pose an insuperable challenge.

HOW TO ESTIMATE YOUR GEAR PERSONALITY

An approximate test to "guesstimate" your gear personality is to see how you react to lying in a quiet, dark room with your eyes closed for twenty minutes either late in the morning or late in the afternoon (outside your post-lunch dip window — *see page 60*).[11] You shouldn't be sleep-deprived or especially tired for any reason. It is also best to resist caffeine beforehand.

- If you fall asleep *very* quickly, long before the twenty minutes are over, your gear personality likely lies at the stiffer end of normal.

- If you are no drowsier at the end of the twenty minutes than you were at the beginning, your gear personality may be at the springier end of normal. Many people with insomnia tend to stay lodged at a high gear all the time. When they do this test, they stay unflinchingly alert and wired.[12]
- Most of us will feel drowsy by the twenty-minute mark. If this is the case for you, your gear personality falls somewhere between the extremes of the spectrum.

It is best to do this test multiple times over several days to cancel out confounding factors on any particular day.

The Hyperefficient Way to Work

n parts of the world where populations pursue a preindustrial way of life, rhythms of the mind seem to trail the rhythms of the body, and both echo the rhythms of the natural world. Nature decides when to work, the body initiates the work, and the mind follows in lockstep.

The goatherds of Seulo and knowledge workers from any bustling city today share the same blue dot network, yet urban knowledge workers find it much harder to maneuver it and adjust their gears. The goatherds wake up at the same time every morning, brimming with vitality, their gears sliding easily into setting 2. For many knowledge workers, however, a buzzing activity tracker, Alexa, a Lumie lamp, and four espressos summon barely enough energy to crawl out of bed. Once their exhausting morning work is done, the Seulo goatherds instantly snap into a gear 1 mode of slow existence. Knowledge workers, by comparison, continue their mental sprint all day and struggle to unwind long after midnight.

Why is one population able to deftly change gears while the other struggles? The answer may lie buried in the long course of our evolution. We never had to change our mental gears alone: we were helped by the rhythms around us.

To imagine how this might play out, think of the human

mind at work sitting on the top of a pyramid of rhythms. The pyramid has three layers. The lowest layer is the rhythm of the natural world: as day breaks, life begins, and when night falls, it winds down. The middle layer is the rhythm of your body as you react to the rhythm of the world: when the sun rises, your autonomic nervous system prepares you for action; when it sets, you slow down. The top layer of the pyramid is the rhythm of your mind. For much of our past, the body's rhythms would set the context for the mind's rhythms: for instance, slow thoughts would accompany a rambling foraging expedition and hunters would have to think quickly while running after prey. The rhythm of the bottom layer percolates all the way up to the top, with each layer casting a subtle ripple on the layer above. Changing mental gears becomes easier if you can ride these ripples, much as a surfer rides an ocean's waves.

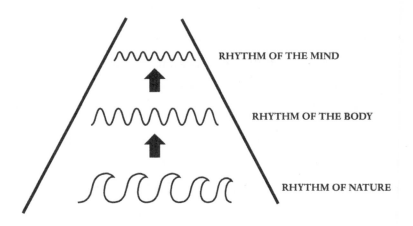

RHYTHM OF THE MIND

RHYTHM OF THE BODY

RHYTHM OF NATURE

The Seulo goatherds seem to do this: they align their mental activity with physical activity and both with the time of day. Knowledge workers, on the other hand, swim against the tide.

In this section, I will reveal how this pyramid can be harnessed to improve mental efficiency across different kinds of knowledge work.

4

THE RHYTHMS OF THE WORLD

"Mankind has got to get back to the rhythm of the cosmos."

—D. H. LAWRENCE[1]

If you sat on a desert island determined to do nothing for a week, your gears would refuse to sit still with you. They would rise at daybreak, continue upward until noon, pause for a dip, climb again, and then finally descend late in the evening.

Every day, the sun's rise and fall elicits at least two rhythms in your body's physiology. One runs for a period of twenty-four hours, and another for a period that is half as long. If you align yourself with these rhythms while you work and ride them just as a surfer rides an ocean's waves, these biorhythms can help tune your gear for you.

Your twenty-four-hour biorhythm

You might remember that your blue dots are your brain's hub for the neurotransmitter *norepinephrine*. In 1976, researchers in Bethesda, Maryland, discovered a twenty-four-hour rhythm in norepinephrine levels in the brains of monkeys.[2,3] Levels were lowest at night and rose over the course of the morning, dropping again in the evening to a nocturnal nadir. To confirm that this pattern wasn't unique to monkeys, the researchers repeated a less invasive version of the experiment on humans and found a similar result. Your blue dot network works to a twenty-four-hour rhythm: it increases in activity over the course of the morning and slows down as you wind down at night.

Your twelve-hour biorhythm

Some research suggests that a *second* biorhythm, a *twelve-hour cycle,* may be nested within your twenty-four-hour rhythm. This cycle prompts you to want to sleep every twelve hours (though the urge is stronger at midnight than at noon). At midday, this pulls you into what is known as a post-lunch dip: a brief period when you feel sluggish and your gear climbs down.[4] You might think that this post-lunch dip in energy levels is the result of eating, but it occurs even when you skip lunch.[5,6] For most people, the dip dissipates later in the afternoon, when your alertness returns.[7,8]

One way to deal with the post-lunch dip is by taking an afternoon nap, for even a short nap in this time bracket has the potential to improve mental performance for up to two hours afterward.[9] The best time for the nap is soon after noon. Alternatively, a preemptive nap at lunchtime — *before* the post-lunch dip sets in — can offset and prevent the dip.[10] Fitting a rejuvenating post-lunch nap into your day can help you sustain mental stamina late into the evening.

Napping removes the fatigue you have accumulated until you nap, and this calls for a caveat. Since accumulated fatigue is one of the factors that drives sleep, you have to clock up enough fatigue between waking up from your nap and going to bed at night to avoid disrupting your nocturnal sleep. Interestingly, if you're having trouble with sleeping at night after napping too much, then learning something right before bedtime — as long as the process doesn't excite you enough to raise your gear — can help you sleep better.[11]

I will explore naps in greater detail later.

Putting the rhythms into practice

Over the course of the day, the trajectory of your gears seems to mirror the arc of the sun. You awaken to the red hues of sunrise in gear 1, and then, as the sun climbs higher, you enter gear 2 territory. At midday, your twelve-hour cycle pulls you into a post-lunch dip, so you feel sluggish and briefly descend into gear 1. This sluggishness gradually dissipates, and your alertness resumes later in the afternoon. As the sun sets, you unwind into gear 1. The precise timing of this rhythm varies, but the trajectory might look something like this:

WHEN TO DO CREATIVE OR FOCUSED WORK ACROSS AN ENTIRE DAY

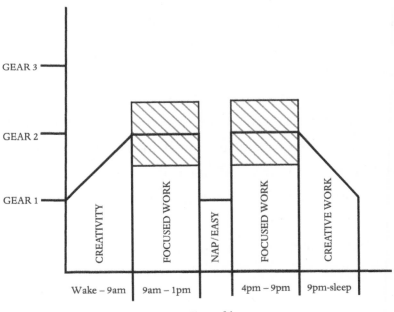

This trajectory provides important clues about the times of day that are most suited to *focused* work and *creative* work.

Focused work (gear 2)

Your ability to stay focused peaks in the late morning and lasts for most of the day, with a brief slump for the post-lunch dip. This would suggest that the best time frames for attention-heavy brainwork happen at *approximately:*[12]

- 9 to 10 a.m. until around 1 to 2 p.m.
- 3 to 4 p.m. until around 8 to 10 p.m.

The time boundaries will vary from one person to another and shift with seasons, climate, and geography. They will also be influenced by how tired you are and how well you slept the night before.

A real-world example of how the time of day affects performance was demonstrated in a study on bank workers who approve loans. After investigating over twenty-five thousand credit loan applications at a major bank, researchers Tobias Baer and Simone Schnall discovered that loans were most likely to be approved in the morning, before 11 a.m. The percentage of loans that were approved sank after 11 and then rose again (though to a lesser degree) later in the afternoon.

It takes more mental energy to approve a loan than to deny one because of the meticulous analysis needed to confirm that a loan will be repaid. Loans were more likely to be approved in the morning and late afternoon, perhaps because these are the hours of peak concentration and focus. If all decisions had been made in the morning, the bank would have made $509,023 in extra revenue in one month![13]

Creative work *(low-energy gear 2)*

Your focused period is *preceded* in the morning by an interval when your mind is ascending from gear 1 into 2, and it is *succeeded* in the evening by a period when it descends from gear 2 into 1. During both ascent and descent, you transition between being gently awake and brightly alert and enter a low-energy gear 2 state, in which your attention can float and forage for ideas, and you can also narrow your focus to filter what you foraged. This state of mind is most suited to creative thinking and happens:[14]

- From waking until around 9 to 10 a.m.
- From around 8 to 10 p.m. until bedtime

If you take a strong dose of caffeine or use other cues to jerk your mind wide awake immediately upon waking, you will shrink the window of morning creativity. Similarly, if you perform heavy exercise, are provoked into excitement (whether the excitement involves happy or angry emotions), or take caffeine late in the evening, you will shrink the evening window of creativity.

Winding down

You need to be in gear 1 in order to unwind. If you are locked in gear 2 or 3, it feels impossibly challenging to fall asleep, and you shorten your sleep duration with a knock-on effect on your mental productivity the following day. This is partly why working late into the night is a bad strategy. By working those two extra hours this evening, you might lose four hours of more efficient focus or creative work tomorrow. Intense emotions during the day can also contribute to keeping your gear locked in a high setting.

How you deal with sleep deprivation and mental fatigue while at work influences how well you wind down. If you force yourself to keep going when you are tired, your brain copes by cranking its gear up to provide maximum horsepower for its struggling machinery — so much so that it almost "gets stuck" at a high setting. This doesn't just interfere with your ability to stay in gear 2 during the day; it also makes it difficult to wind your gear back down at night when you want to fall asleep.

Your personal clock

The ebb and flow of the twenty-four-hour biorhythm can vary by a small margin from one person to the next.

"Evening" people or "night owls" feel better when they begin and end their days a few hours behind "morning" people or "larks," who have a preference for waking and sleeping especially early.[15]

These dynamics manifest in the way the core temperatures of owls and larks fluctuate with their twenty-four-hour biorhythm. Your body cools down when you sleep and heats up as you awaken. The point at which your body is coolest is used as a "landmark" for your individual twenty-four-hour rhythm, and this landmark is shifted in larks and owls. In one study, larks were coolest at 3:50 a.m., and owls at 6:01 a.m. People who were neither larks nor owls were at their coolest at 5:02 a.m.[16]

Studies estimate that a quarter of people are either larks or owls. Almost everyone else is positioned somewhere in between. Where you fall on the lark-owl spectrum can change with age, situation, or context, and there remains some debate in the research world about whether chronotype (your body's natural inclination to sleep at a certain time) really is hardwired or not. For instance, your social environment, too much blue-light exposure, or long work hours could be responsible for your habit of staying up late every night. Some owls become more lark-like when they hit the calmer life of middle age.

In one experiment[17] in which larks and owls were given simulations of mental work to do at 8 in the morning and

8:30 in the evening, larks performed better in the morning and owls performed better in the evening when their work required thought and precision. The larks also performed well for *longer* when they worked in the morning compared with the evening. The creativity windows occur later in owls and earlier in larks. The post-lunch dip can also vary with chronotype. If you are an owl and wake up at 10 in the morning, your dip may shift to later in the afternoon, whereas if you are a lark who has been up since 5, your dip may happen sooner, even before noon.

If you aren't sure where you are on the lark-owl spectrum, one way to estimate it is by asking yourself questions such as:

- If you were on a desert island with nothing to do, what time would you wake up and when would you want to go to sleep?
- Do you always need an alarm clock to wake up in the morning?
- Do you feel bright and fresh when you climb out of bed in the morning?
- Would you be able to exercise soon after waking in the morning?
- If you had to sit for a tough two-hour exam, what time would you prefer to schedule it?
- Would you struggle to fall asleep at 11 p.m. tonight?

- If you had to wake up at 3 in the morning to catch a plane, would you go to sleep early the night before or catch up on your sleep during or after the flight?
- If you could work only four-hour days, when would you work?

These questions are similar to those used in official questionnaires to measure chronotype. Since this is not an objective measure and your personal rhythm will be influenced by many different things, it should not be used prescriptively. But it will give you a sense of your "morningness-eveningness."

An assembly-line world of work, where everyone is expected to begin at the same time every day, places both owls and larks at a disadvantage. If you're an extreme lark, your window of focus may already be over by the time you begin work at 9 in the morning, whereas if you're an extreme owl, you cannot be productive at work for several hours at the start of each day. Owls continue being productive later into the evening, long after they have left work, while larks often stop being productive long before they leave work. Some researchers have reported that owls may stay awake even later than their natural biorhythms dictate because they have to resort to more caffeine than larks to feel alert in the first part of the day. When morning arrives, these owls have slept very little and need even more caffeine to feel alert enough to work. This vicious circle can put them into a state of chronic sleep deprivation, which compromises their cognitive performance overall.

Whether you are an owl, a lark, or somewhere in between, it may be more productive for you to align your work hours with these rhythms, rather than trying to align them with your work hours.

THE SIGNALS FROM NATURE'S RHYTHMS

Nature is constantly sending you sensory information to help you gauge the time of day. The color spectrum of sunlight changes from morning to night, and as the world wakes up and retreats in slumber, what you can hear changes, too. Light and sound alone can help change your gear, perhaps due to the ubiquitous coordination among time, light, and sound in nature.

Light

One of my most memorable sunrises was at Bayon temple, in Siem Reap, Cambodia. We arrived at a site in the jungle beside the temple at 4 in the morning and waited for the sun to illuminate the haunting giant faces on the temple's façade with its first rays. It was pitch-dark and eerily quiet at first, the silence punctured now and again by piercing animal cries coming from deep within the jungle. Presently a soft whisper of rhythmic monastic chanting drifted from a temple hidden somewhere in the darkness. As it grew louder, a crimson glow emerged over the tree-lined horizon.

The cries of the jungle now gave way to birdsong and the quacking of a mother duck as she led her ducklings toward a pond. The sun's warm glow turned into a bouquet of rich, hibiscus-red rays, and the chanting, having reached a crescendo, ceased. The temple looked so spectacular that even the chanting monks arrived to take photos with their smartphones.

This protracted unfolding of dawn demonstrates how sunrise is a gradual process that cannot be reduced to an LED light switch. If you have slept outdoors and awoken under a clear sky, you will know that a gentle red glow suffuses your eyes before you open them to see a red-orange horizon. If you switch on a blue-dominated LED light first thing in the morning, you miss these red-hued phases. Gradually, as the sun climbs higher, the red-orange gives way to bright white daylight. The evening follows the same pattern, but in reverse. Daylight doesn't turn into darkness in an instant; it transitions gently through intervening phases.

We are now learning how these phases of color induce distinct states of mind, with blue-hued light augmenting alertness and warmer, red-hued light encouraging a calmer state of mind.[18] Cool-hued light can help you stay vigilant and focused if you are tired, whereas soft, warm light is better for creativity. By waking up to a red-orange sunrise, your brain gets the chance to gradually "warm up"; it gently surveys the world in gear 1 before launching into a sharp, bright-eyed, focused mode in gear 2.

Sound

Sound has a peculiar quality. It defines and gives character to a space without being visible. I experience this often when I am waiting for the last Eurostar train of the night to arrive at London's St. Pancras station. At that late hour, the station is abandoned and bare, its magnificent arches echoing the sparse murmurs of late commuters. Suddenly, a soft melody diffuses like mist into the lifeless quietude. As you follow the mist, it grows louder, and leads you to the gates of the Eurostar platform, where some stranger is sitting at a public piano, playing for no one and nothing but the pleasure of playing itself. A small crowd is often gathered around the pianist, who has, with his music, built a room without walls, a room where a celebration continues into the night, while the rest of the station sleeps.

Music — and sounds in general — moulds your perception of reality by coaxing your mind into different states. Hence its ubiquitous use as a tool for "setting the scene" in films and plays, and for helping people reach deep, meditative states in temples and monasteries. It is a useful tool for entering gear 2, staying in gear 2, and winding down from gear 2 at the end of the day.

Generally, the higher the frequency, tempo, and amplitude (volume) of the sounds in your environment—including speech—the more they raise your gear.[19] This dynamic is utilized by morning shows on television and radio, whose hosts speak fast, loudly, and often with higher-pitched voices in an attempt to rouse sleepy viewers out of a gear 1 state so they pay attention.

Conversely, lower, slower, and softer sounds—including speech—lower your gear, and this principle is used in spas, dental practices, and hotel lobbies. In fast-paced or high-pressure situations in which you feel yourself sliding up to gear 3, slow, rhythmic music lowers you into gear 2.[20] In a study on dart players, listening to relaxing music before a game improved performance.

Sounds influence your brain partly by "entrainment," a process whereby the electrical oscillations in your brain latch on to and follow the rhythmic oscillations in your environment, and partly through your brain's interpretation of the sounds, including their emotional meaning.

The effects of any kind of music or sounds on your focus and productivity depend on the kind of music or sounds you listen to and the gear you are currently in.

Fast-tempo music or general loud noise can help you

get into the right frame of mind if you are in gear 1 but can do the opposite if you are already in gear 2. For example, if you are bored and falling asleep at your desk, relocating to a noisier environment, like a crowded café, can improve your focus and help you work better. But if you are already in peak gear 2, moving to a place with more noise will take you out of your zone of focus and harm productivity. Background music of any kind that has lyrics interferes with focused work because the lyrics present a stream of information that competes with what you are working on and seduces your attention away.[21]

If you have difficulty focusing in a sea of clicking mice and keyboards or other distractions, listening to white noise can help. White noise has a bidirectional effect on your state of mind. If your background is noisy and distracting with beeping alarms, ringing phones, and chatter, white noise drowns out those distractions and helps you focus. If your background is so quiet that you feel yourself drifting into somnolence, white noise alerts you just enough to stay in gear 2. White noise is particularly helpful in open-office-plan environments.

Any acute noise that is louder than the baseline of the ambient noise around you — beeping alarms, people chattering, phones ringing, et cetera — raises your gear. There is a limit to the volume of ambient noise beyond which mental performance becomes impaired, but this limit can vary with the type of noise and its pitch, with different researchers citing levels from 70 decibels (about as loud as a running washing machine) to 95 decibels (about the volume of a motorcycle engine).[22] Your gear personality can influence how sensitive you are to different levels of noise. If you have extremely springy gears, you might react strongly to a level of noise that someone with especially stiff gears would barely notice.[23]

The Rhythms of the World

SIMPLE TIPS TO RAISE OR LOWER YOUR GEAR

	RAISE	**LOWER**
With sound	• Add white noise if your surroundings are too quiet • Add background music and increase the pitch/tempo/volume	• Lower background noise (or add white noise if you cannot) • Lower the pitch/tempo/volume of background music
With light	Brighter, with cold (blue) hues	Dimmer, with warm (red) hues
With your surroundings	Move to a space where people are moving and speaking faster and louder around you	Move to a place where people are moving and speaking slowly and softly around you

5

THE RHYTHMS OF THE BODY

Jumping from boulder to boulder and never fall-
ing, with a heavy pack, is easier than it sounds;
you just can't fall when you get into the rhythm
of the dance.

—JACK KEROUAC[1]

For much of our past, brainwork and muscle work were close partners. One faithfully accompanied the other, the mind complementing the body and the body doing the same to the mind, through work, rest, and play. This mutuality is shown to spectacular effect in the BBC documentary series *The Life of Mammals*, featuring Sir David Attenborough.[2] In the episode "Food for Thought," Attenborough follows the San people of the Kalahari while they hunt kudu. The hunters begin by tracking the kudu over the course of some hours. They move at a moderate pace and enter a trancelike state of concentration. Once the bull separates from the rest of the herd, one hunter, Karoha, gives chase. Suddenly, the bull's tracks vanish, and Karoha pauses to figure out what happened. He stands still, then walks slowly while trying to trace the bull's trajectory in his mind. The hunter's thoughts and movements, the pace of his mind and the speed of his body, faithfully mirror each other at every phase of the hunt.

We began divorcing the body from the mind at scale in the era of assembly lines, when the body worked while the mind slept. More recently, knowledge work has taken this to the opposite extreme: the mind works while the body sleeps. This bifurcation has robbed the mind of the plethora of signals that came from the body and helped to tune its gear. Without these nudges, the brain must work harder at getting into the right state at the right time, and it inevitably wanders into the wrong one.

In this section, I will describe two routes through which your body's rhythms influence how your mind works: through the movements of your body and the movements of your eyes.

THE MOVEMENTS OF YOUR BODY

Attenborough's documentary demonstrates how an alliance between the body and the mind dictated much of preindustrial existence. This alliance, while somewhat subdued, persists in the life of a modern-day knowledge worker. It explains why when your mind panics, your body does, too. You can manoeuvre the body to modulate the mind by harnessing aspects of this dialogue.

One conduit through which the state of your body alters the state of your mind involves an extensive tree of nerve connections called your *autonomic nervous system*. This system meticulously tracks everything you do and continuously tunes your body to the demands of the world using its two big arms: one arm (known as sympathetic) *accelerates* and the other (known as parasympathetic) *decelerates* your body's power and pace. At every passing moment, the world demands slightly more or less action, and the two arms adjust you accordingly. Consider something as commonplace as standing and sitting. When you stand from sitting, your body must work slightly harder against gravity to push blood into your brain, so the autonomic system accelerates; when you sit down again, it doesn't need to work quite as hard, so the system decelerates.

Barring some exceptions, your mental gear marches in step with your autonomic nervous system. This is why it is impossible to daydream in the middle of a 100-metre sprint (when you are in gear 3) and why stretching the body relaxes the mind (by placing you in gear 1).

Shifting your autonomic state also shifts your gear. Since your autonomic state changes when the world demands more or less action from your body, one way to shift your gear is by giving your body more or less physical work.

General Exercise

Any exercise that makes your body work harder than your resting baseline raises your gear.[3] The more intense the exercise and the longer its duration, the more potent its gear-raising effect.

Your gear doesn't shift down the moment you stop exercising — it stays raised for a short while afterward. This is why you feel more alert after exercising if you were feeling sluggish before. Moderate exercise for around half an hour has measurable benefits on how fast you think and on general mental performance for at least half an hour afterward.[4]

This effect of exercise may be especially significant for those suffering from ADHD and other attention disorders. Earlier, I described how some studies have revealed a link between attention deficit disorders and compromised gear regulation, and exercise can be leveraged to make it easier to enter a gear 2 state of focused attention. In one study on adults with ADHD, thirty minutes of cycling at moderate intensity boosted attention and processing speed in mental work begun within forty minutes afterward,[5] and a similar protocol helped ADHD sufferers to resist distractions and impulsive reactions.[6]

Muscle Contraction

In 1943, a University of Chicago study found that students could improve their performance on arithmetic and reasoning problems *simply by contracting their muscles*. Subsequent research has confirmed that your autonomic system generally tilts in the direction of acceleration when you contract your muscles, and your gear shifts up in chorus.[7] This means you can make yourself feel more alert and raise your gear almost instantly, at least to some degree, simply by tensing your muscles.

In a 2020 study, Mara Mather and her team at the University of Southern California showed that periodically squeezing a "stress ball" with one's hands for eighteen seconds, relaxing the hands for a minute, and then repeating the process several times improved performance on a concentration- and focus-based mental test taken soon afterward.[8]

Just as contracting your muscles shifts your autonomic system toward acceleration, *relaxing* your muscles after contracting them or *releasing* them after a stretch shifts it

AmE toward *deceleration*. This forms the basis of hatha yoga (which incorporates both of these mechanisms) and progressive muscle relaxation (PMR), or body scanning (which involves contracting and then relaxing muscles in an ordered way).

Body Temperature

Your autonomic nervous system regulates your body's temperature, and it jolts into action when something threatens to push your core temperature up or down. This is why your heart rate rises in a hot sauna and when you first jump under an ice-cold shower. Although a sauna and an ice shower shift your temperature in opposite directions, both trigger a sympathetic corrective response from your autonomic nervous system, which raises your gear and elevates norepinephrine levels. Any gentle "shock" to your body temperature will raise your gear, and both hot and cold showers will shake you into gear 2 from a lethargic gear 1 first thing in the morning.

A hot bath has a dichotomous effect when taken before bedtime. The process of cooling down after taking the bath can lower your gear and help you fall asleep.

The Way You Breathe

In 1976, Jacques Mayol became the first person in recorded history to dive 100 metres under the sea without coming up for air. At that depth, the slightest rise in mental excitability can amplify your need for more oxygen — and tip you over the gossamer-fine line that separates life from death. One technique Mayol used to train himself for this tremendous feat was yogic breathing.

Slow breathing, and prolonged exhalations in particular, lowers your gear. A few years ago, researchers at Stanford University uncovered a clue that may explain why this happens: a direct nerve highway links your blue dot network and the part of the brain involved in setting the pace of breathing, called the pre-Bötzinger complex.[9] When you slow one down, it slows the other down, too.

Breathing at a rate of about ten seconds per breath, approximating to a cycle of five seconds inhalation and five seconds exhalation — the frequency used in pranayama techniques — can stimulate the *vagus nerve,* one of the "calming" branches of the autonomic nervous system.[10] Striking new research findings from Mara Mather's lab at the University of Southern California hint at the possibility that stimulating the vagus nerve in this way directly shifts the blue dot network into a gear 2 pattern of firing. Practically, this implies that breathing at a pace of about five or six breaths per minute can be used as a tool to remain in gear 2 or rapidly return to gear 2 at critical moments when anxiety or stress pushes you into gear 3.[11]

The peculiar state of "calm alertness"

The mental states of being alert and calm are inherently contradictory at their extremes: the more relaxed you are, the less vigilant you become, and vice versa. Yet there is a sweet spot somewhere in the middle where you can be *both* overwhelmingly calm and intensely alert at the same time. One way to reach this state in theory would be by activating both the *accelerate* and the *decelerate* arms of the autonomic nervous system *simultaneously*. There is some—albeit anecdotal and early—evidence of specific activities ticking this box.

Cold-water bathing

One such activity is cold-water immersion. When you immerse yourself in water, two almost opposite reactions take place at the same time. We react to immersing the face in cold water with a reflex known as the dive reflex, which slows down our physiology by activating the calming branch of the autonomic nervous system. The shock of the cold water, however, *also* triggers the accelerating branch of the autonomic nervous system, resulting in an alert, fight-or-flight response. Some researchers have speculated that this gives rise to a kind of "autonomic confusion" in which both the alerting and calming arms of the autonomic nervous system are maximally turned on at once.[12] This may partly explain why many people find that this activity leaves them in a state of peak focus and feeling profoundly tranquil, both at the same time.

Hot yoga

Hot yoga (such as Bikram yoga) can have a similar effect. While hatha yoga *decelerates* the autonomic nervous system through a series of stretch-and-release and contraction-and-release movements reinforced by slow, controlled breathing, the hot environment (the temperature in Bikram yoga studios is set at 42 degrees Celsius) stokes the *accelerate* branch of the system, which is hard at work trying to keep your body temperature down. I feel both intensely alert and deeply relaxed after each session — an odd combination that I don't experience after any other activity.

THE MOVEMENT OF YOUR EYES

You can often gauge a person's state of mind just by watching their eyes.

If your hunter ancestor found himself surrounded by ten pairs of lion eyes one sunny morning, his gaze would have likely darted erratically from one pair of eyes to the other. This behaviour would have sent a danger signal to his blue dot network and raised his gear. If, on a different morning, this same ancestor went foraging for food and discovered a peculiar shrub he had never encountered before, his gaze might have lingered on the shrub for a long time while he studied it, his mind shifting into gear 2 as he zoomed in to its details.

The patterns of your gaze have a close relationship with your gear network because how you look at things affects how you pay attention to them. You can sometimes shift into a gear 2 state of unwavering focus or to a low-energy or high-energy gear 2 state of creativity by playing with your gaze.

Narrowing your gaze

Your eyes dart around when you are anxious or excited. When you are immersed in concentration, however, they stay very still.[13] Using this dynamic, actively forcing your eyes to stay still by holding your gaze on a small, focal target enhances your focus and anchors you into gear 2. This technique, known as Quiet Eye, is used in an array of high-tension situations, from competitive archery and basketball to target practice and surgery, when you want to dial your gear down and tip back from gear 3 into gear 2. As a practice, this method isn't new; meditators have long used it to enter deep states of concentration.

Surgeons, snipers, hunters, and even tennis players who fix their gaze to within 3 degrees of a target and hold it for just 100 milliseconds before taking aim perform better than those who don't.[14] As W. Timothy Gallwey writes in *The Inner Game of Tennis*, "I have found the most effective way to deepen concentration through sight is to focus on something subtle, not easily perceived. It's easy to see the ball, but not so easy to notice the exact pattern made by its seams as it spins...The pattern...is so subtle, it tends to engross the mind completely."[15]

When surgeons, particularly microsurgeons such as ophthalmologists, neurosurgeons, and vascular surgeons, operate, it is important that their hands do not shake. If their gears overshoot into gear 3, it affects fine control over their hands and their ability to tie knots. The Quiet Eye technique seems to help mitigate this.[16]

If you are already in focusing mode, then forcing your eyes to stay on a target won't make much difference, but if you're anxious and feel your self-control slipping away, the practice can help restore your concentration.

Widening your gaze

When you concentrate hard on something, you tend not to notice things going on in the periphery. Some research suggests that forcing yourself to pay attention to a wider area in your visual field can nudge you into a more creative mindset.[17] In some ways, this is the opposite of the Quiet Eye technique. A reason why this might happen is that when your brain registers information in the peripheries of your visual field, it does the same for the information in the peripheries of your mind. If you widen your mental aperture so it can keep more ideas in focus at the same time, you are more likely to spot connections you were not aware of before.

Some research even suggests that your facial expressions can color your creativity. In one experiment, participants who were instructed to keep their eyebrows raised while trying to come up with novel uses for a pair of scissors generated many more original ideas than participants who had been instructed to do so while frowning! The researchers speculate that frowning narrowed their scope of attention and raised eyebrows widened it. It is tempting to conjecture that this is why we raise our eyes in wonder or surprise: it might make us think more creatively and flexibly in order to navigate ourselves around a new or unexpected situation.

Detaching your gaze

Sometimes focusing too intensely on your work is unhelpful. You become so entangled in what you are doing that you miss the forest for the trees. Taking a step back can help you gain perspective and notice details that you could not see before.

One way to loosen focus is to detach your gaze, and the simplest way to do this is by briefly closing your eyes.

Your eyelids are like curtains: closing them blocks out all magnets of attention in the outside world. When your eyes are shut, your attention drifts AmE inward, into your mind, and it becomes easier to dig out an evasive memory or solve something in your head. This is why people often close their eyes when they are immersed in deep thought or on their way to a creative breakthrough.[18] If you are agitated by a problem at work, you can lower your consternation by imagining it in your head with closed eyes instead of looking at it on a screen. This dims the problem's intensity and provides perspective.

A second way to unglue your attention is by "looking at nothing." Blank walls and empty spaces make the outer world too boring to pay attention to, so your attention drifts and wanders.

Any situation that lacks an anchor for your attention can help your mind to drift. Dull meetings, long showers, and washing the dishes are some examples. When you walk or jog, the world around you is in constant motion; your attention can't affix itself to any target and is set loose to take a stroll and stumble upon *a-ha!* moments of insight.

Briefly detaching your attention from what lies in front of you can help in three ways:

- It wipes your mental slate clean so you can disentangle yourself from a problem, see it with fresh eyes, and gain perspective.
- It gives your mind the chance to rest and recharge.
- It frees your attention to be reallocated to another task, or to wander for creative ideas.

The last point is especially relevant in a work environment where you have to frequently interrupt what you're doing and switch your attention to something else. If your attention is deeply entrenched in what you are doing, it feels difficult to shift it instantly onto a new target. But if you loosen your attention *before* you try to move it, it shifts much more easily. Think of it as changing gears in a manual car. You need to bring the stick into the neutral middle before you move it to a new position.

Detaching your gaze without leaving gear 2

When you unglue your attention for too long, whether by closing your eyes or "looking at nothing," you might find yourself relaxing to the point where you drift into gear 1 and temporarily lose the ability to focus. One way to detach your focus *without* drifting into gear 1 is to use a cue to stay alert while your mind drifts. One of the most time-honoured and effective strategies for achieving this is *walking*.

Walking creates a unique mental state: it lets your attention float while keeping you in gear 2. It achieves this by virtue of three features:[19]

First, the act of walking keeps you alert, but not *too* alert. Running can tilt you into gear 3, which blocks creative ideas, but (depending on your fitness level), walking does not normally push you beyond gear 2.[20] This may explain why there are more anecdotes of people getting new ideas while walking than while running.

Second, while you walk, the world moves past you slowly enough for you to gaze at it, but too fast for you to fix your gaze upon it. You pay attention to it but it cannot hold your attention. When your floating attention drifts into the territory of your mind, you can think about a problem in your head, but you cannot become too entrenched in the problem because you still have to direct your attention outwards, into the mechanics of moving, from time to time. This forces you to keep a wide-angled view of any problem you are trying to solve and repeatedly refreshes your perspective.

Third, when you move relative to the world around you, your brain processes more visual information coming from the periphery of your visual field, which can widen the beam of your attention.[21]

Where you walk makes a difference. Walking on a treadmill where the world around you is generally fixed doesn't expand your attention field like walking on a normal surface. *How you walk* also matters. The more restrained you are, the narrower your focus becomes. Walking in a restrained way by carefully following a prescribed route doesn't improve creative thinking as well as walking in a free and unrestrained way.[22]

Curiously, this "restraint" effect applies to movement of the body in general. If you are sitting in a restrained way and have to hold your gaze, your creative scores will be lower than if you are sitting the way you want to sit. It even applies to hand movements. Moving the hands freely in a fluid way (doodling, for example) increases scores on creativity tests, whereas drawing straight lines does not.[23]

STIMULANTS

Tea and coffee are among the most popular tools we use to feel alert and focused. They contain caffeine, which raises your gear.

Caffeine

The formal practice of taking a coffee break is said to have been born when a tie-manufacturing company in Denver, Colorado, discovered that when workers took a ten- to fifteen-minute break and drank coffee in the morning and afternoon, they worked faster and increased the company's revenue.[24] It is no coincidence that the rise in coffee's popularity in the nineteenth and twentieth centuries mirrored the pace of the industrial revolution, as author Erin Meister points out in *New York City Coffee: A Caffeinated History.*[25] Once assembly-line manual work became assembly-line-style *mental* work, coffee continued to turn the cogs of knowledge work factories, as evidenced by the ubiquity of coffee bars in knowledge work hubs around the world.

As a stimulant, caffeine raises your gear and can compound the effect of other gear-raising events, such as stressful experiences and exercise. Taking caffeine beforehand can induce a bigger autonomic and gear swing during moderate exercise and prolong the window during which your gear stays raised afterwards.[26,27] This caffeine-exercise dialogue can be leveraged to delay a boredom-induced collapse into gear 1 or simply to get into the right mental state for focused work.

Although caffeine helps you enter a focused state of mind *if you are in a low gear to begin with,* if you are already in gear 2 or gear 3, caffeine will make it *more* difficult to concentrate, and your ability to filter the signal from the noise will be impaired. In this state, you will be alert but distracted, and you might work faster, but you will make more mistakes, especially in work that requires careful thought.[28] These effects can be unpredictable if you are a seasoned caffeine drinker.

Caffeine's effect on your ability to think creatively and solve problems is nuanced. It helps you converge disparate elements into one explanation or solution, but it does not help you think outside the box or come up with original ideas, for which your mind needs to wander. In one experiment involving moderate daily caffeine drinkers, 200 milligrams of caffeine (approximately equal to one 12-ounce cup of coffee) significantly enhanced convergent problem-solving ability but did not improve divergent thinking or working memory.[29] Caffeine applies an upward push to your gear, which may make it more difficult to climb down into a less energetic state of mind.

Caffeine is a useful tool in professions where sleep deprivation is common and the risk of waning alertness is high. When ten Special Forces personnel took a 200-milligram dose of caffeine four times a day (twice each evening and twice each morning) over a three-day period during which they did not sleep at night and survived on just a four-hour nap every afternoon, they had faster reaction times and were more vigilant and alert than personnel who took a placebo. While this amount of caffeine consumption may help save lives on the battlefield, it is not recommended for the typical knowledge worker.

Other Stimulants

Most stimulant drugs, like the kind prescribed to those with attention disorders, raise your gear. By reducing the effort needed to elevate your gear, they can make focused work feel easier if your gear is drawn down through fatigue, sleep deprivation, or boredom. But whereas you may feel yourself thinking and concentrating better, the *quality* of your thoughts aided by stimulants may not be optimal. Similarly, while they may make you feel more alert and less tired, at least temporarily, stimulants don't undo the chemistry of fatigue. They don't rejuvenate your mental resources if you are exhausted; they raise your gear so your performance *dissociates* from fatigue. This is how your gear stays high despite fatigue.

By overriding your brain's reflex to pull you down into gear 1 when you are tired, stimulant use can block your descent to gear 1 at the end of the day, which is why insomnia is a known side effect.

The Rhythms of the Body

SIMPLE TIPS TO CALM DOWN OR BE MORE ALERT

To be more alert (raise your gear)

- Up to 30 minutes of mild (*not exhausting*) exercise
- A mild to moderately intense session of lifting weights or tensing your muscles for a few seconds where you are
- A hot *OR* cold shower
- Caffeine in moderation

To calm down (lower your gear)

- Slow, deep breaths at approximately six breaths a minute
- Fix your gaze onto a narrow target
- Detach attention from the source of your distress:
  - If your distress comes from the work in front of you, get up and take a walk
  - If your distress comes from thoughts inside your head, immerse yourself in an absorbing activity

6

THE RHYTHMS OF THE MIND

The king... shall divide the day and night, each
into eight periods of one and a half hours, and
perform his duties.

—KAUTILYA'S *THE ARTHASHASTRA*, 300 BC[1]

In an essay published in *The Economist* in 1955,[2] British historian Cyril Northcote Parkinson noted how work often "expands so as to fill the time available for its completion." Writing a few lines on a postcard, he observed, can take an entire day if one has the entire day to write it in. Various factors, from perfectionism to procrastination and laziness to lassitude, contribute to this phenomenon, which has come to be known as Parkinson's Law. When it applies to why you don't get more done by working for three hours straight than by working for two, however, Parkinson's Law may be explained by the *rhythmic cycle of your mind.*

hyperefficient

Your brain swings itself like a pendulum across different states while you sleep. Every ninety minutes or so, you cycle between deep, dreamless sleep, when your mind is motionless but your body is not, and dream-filled REM sleep, when your body is motionless but your mind is not. (Your muscle tone falls during REM sleep. Those who sleepwalk tend to experience the phenomenon during deep sleep.) In the 1960s, American physiologist Nathaniel Kleitman and others speculated that this ninety-minute pendulum is also in motion while you are awake, driving you through subtly rising and falling states of alertness. Kleitman called this cycle the basic rest-activity cycle, or BRAC, and surmised that the cycle is barely perceptible in a healthy person because it is easy to override.[3]

The subtlety of the cycle makes it challenging to study in a real-world context, but in 2017, a Mexico-based research team tried a new approach. They used wearable technology to track a group of young adults with short-term sleep deprivation as they went about their day and compared the results to a control group who were not sleep-deprived.[4] While those in the control group were comfortable doing different activities for variable lengths of time across a whole day, the sleep-deprived individuals seemed less able to continue an activity for longer than about ninety minutes without some kind of interruption. It was as if the ninety-minute cycle described by Kleitman emerged when people did not have the resources to override it, in this case because they were sleep-deprived.

Working on something continuously without a break for more than about ninety minutes feels tiring for most of us. When people resume intense mental work after taking a refreshing break, they stay alert and focused for about ninety minutes before returning to the same state of fatigue as before they took the break.[5] As mental tiredness kicks in, your mind drifts and your performance starts to sink, but your brain tries to compensate for its waning performance. A Florida-based research team has identified at least one brain region that may be involved in this compensatory mechanism; it swings into action when other brain regions start to slow down.[6]

We have designed certain features of our lifestyles with an almost intuitive assumption that attention wanes in ninety-minute cycles. Early Hollywood films were often cut at the ninety-minute mark, resulting in American audiences viewing a heavily truncated version of Fritz Lang's 1927 masterpiece, *Metropolis*. Football and rugby matches are approximately ninety minutes long. Play breaks at many schools bifurcate the mornings and afternoons into roughly ninety-minute segments, as do tea and coffee breaks in workplaces. In a 2003 paper, Poland-based researchers conjectured that a ninety-minute rhythm is wired into our brains to protect the mind from information overload and give it a chance, at regular intervals, to wander and process what it has imbibed.[7]

HOW TO STRUCTURE A WORK SESSION

Tying these threads together, it would make sense to schedule work sessions that are around ninety minutes long, though this timing can be modified depending on how tired you feel, the time of day, and the kind of work you are doing. (If you are working later in the day with depleted mental resources, a sixty-minute session may be better.) By the time you hit the end of your session, your mind may already be drifting into gear 1 mode as it winds down to catch its breath. This is the time for you to take a break and recharge.

Because your mind is freshest at the start of each session, you could aim to do 80 percent of the most complex, mentally demanding work in the first 20 percent of your sixty- to ninety-minute work session, and the least demanding work in the latter part of each session. If you have multiple tasks to get done during a single session, it helps to rank them from taxing to easy and do them in that order. If your hardest task takes longer than about twenty minutes and you find your pace starting to sag, it might help to put it aside until the start of your next work session, when your mind is refreshed.

On this basis, each work session would look something like this:

- Perform the hardest tasks in the first twenty minutes.
- Do slower, easier work for the remaining forty to seventy minutes.
- Take a break for ten minutes.
- Repeat.

HOW TO STRUCTURE A 90 MINUTE WORK SESSION

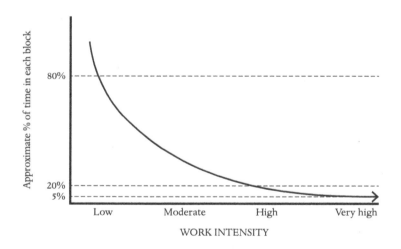

In general, you should limit intense mental work to no more than four hours a day in total. When you work intensely for longer, your mind is left so tired that it cannot recover after a night of rest and your fatigue is carried over into the next day, adding to that day's load.[8]

Pure, creative work needs a different approach — it is best done without any limits or structure. If your mind is deep in creative ideation, then imposing a strict schedule can interrupt its thinking. Take the example of KR and ST: KR, an engineer in an advertising firm, uses the "pomodoro technique" to remind himself to stop for a break every sixty minutes, and a tomato-shaped kitchen-timer-like clock buzzes to tell him when his break is due. His colleague, ST, sits at the workstation next to him and designs user interfaces. ST relies on creative thinking, and his mind needs to be in a low-energy gear 2 state to keep churning out ideas. As he watches KR's clock, ST gets an uneasy feeling of time rushing by. When the alarm rings at the end of each cycle, ST's gear shoots up and his mind snaps out of its creative galaxy. Any *a-ha!* moment he was about to experience is immediately decimated.

Regardless of the type of work you are doing or how you structure your work sessions, the important feature is the waxing and waning, the power law-like pattern of working. This approach can be stretched macroscopically, too. When you have a really intense day at work, sandwich it with a couple of light days. If you are a manager, structuring projects this way can help protect your team from burnout.

WHEN THE MENTAL MUSCLE GETS TIRED

If you are a Seulo goatherd and have, with considerable mental effort, just figured out why your goat has been limping lately, you might sit down for a moment to recover your mental stamina. But if you are a knowledge worker in a large corporation and it is 11 in the morning, the urge to take a break after an equally intense bout of mental gymnastics would barely register. The notion that you should stop working when you begin to feel mentally tired is almost inconceivable during "working hours" in most knowledge workspaces. One reason is we can't *see* a reason to stop. If you are lifting heavy weights, everyone can see when you are perspiring, breathing heavily, and wincing with physical pain. None of these signs accompany mental fatigue.

When your work demands mental "heavy lifting," a network of brain cells in your prefrontal cortex (called the "cognitive control network" or CCN) buzzes with activity. As your mind gradually tires from the "heavy lifting," the network's activity slows down. There are different theories as to why this happens. It is possible that intense mental work exhausts resources so the brain cells can no longer work properly. Another theory, proposed by researcher Clay Holroyd in 2016, posits that the harder the brain cells in the CCN work, the more "waste" they produce as a by-product.[9] As these waste products accumulate, they signal to the network to slow down, which gives the brain's clearing process a chance to catch up. This feedback loop prevents too much waste from piling up and causing harm. Through the lens of this theory, mental fatigue is a sign of toxic waste building up at a rate faster than it is being cleared away. If this theory is proven to be true, then every time you override the urge to stop and take a break, you are forcing your brain to endure a pileup of toxic garbage!

A technique that involves analysing the brain's electrical activity using graph theory has shed further light on the inner workings of a tired mind. When bits of information dart across the brain, they follow the most efficient routes available. If you think of the information as cars and the routes as highways, the cars don't always stick to the same highways when getting from point A to point B. If one highway is congested, many cars will avoid it and take a different route. Or if one highway has tolls and the other doesn't, many cars will choose the cheaper route. If you study these data "traffic routes" in the brain in real time, you can get a feel for how these calculations change as your brain grows tired. When, in 2017, a research group based in Singapore did just that, they discovered that performing intense mental work made traffic routes less efficient over time.[10] If we wind Holroyd's theory into this, we could conjecture that when the most optimal routes become compromised by too much toxic waste, the brain reorganizes itself and uses alternative and less efficient routes to relay information.

Mental fatigue is a signal that your brain wants — and needs — an opportunity to replenish resources or get rid of accumulating waste that has built up from intense mental work. These compromised resources are preventing your brain from working efficiently, so it needs *more effort* to do the same work, and the work feels harder to do.

The built-in mechanism to deal with this is to *slow down*. When you feel yourself disengaging from your work because of mental tiredness, that feeling is your brain sliding itself down into gear 1 to move at a slower pace. Brain scans show that just as the CCN winds down with fatigue, another brain region, the DMN, steps up. The DMN, or the default mode network, encourages the mental state of gentle daydreaming, and it comes alive in gear 1. Gear 1 provides a respite from intense work and lets the brain recover.

Motivation is another factor in the fatigue equation. Motivation raises your gear, and its absence lowers it. When you become tired from doing mental work, one of the first things you might notice is that you don't want to carry on working; it becomes increasingly difficult to recognize that something is interesting or enjoyable when you are in an exhausted or a distracted frame of mind. Dwindling motivation pushes you down into gear 1, where you feel less willing and able to invest the kind of effort you were putting in when your mind was fresh.

For much of our past, few things other than a pressing need for food or safety would have forced us to keep doing what we were doing if we felt no motivation for it. Today we don't stop when we grow tired or unmotivated; instead we carry on with extra effort. When you keep going despite fatigue, your brain needs more help, so it mobilizes extra resources by going into gear 3. It is similar to the way you have to push down hard on your pedal when you cycle uphill.

- *Mental fatigue draws you down into gear 1.*
- *Attempts to override or push through mental fatigue lift you into gear 3.*

You can do mindless or automatic work well in gear 3, but not work that needs thought or concentration. This is what happens when you feel "tired and wired": your mind is buzzing, but it is also too exhausted to solve a math puzzle. In this state you struggle to focus, ignore distractions, and overrule unwise responses. And the more you try to push through, the more your performance suffers.

Relying on gear 3 as a kind of crutch to keep you going through hours or days of intense mental work results in a buildup of fatigue that can take weeks to recover from.

One small study found that students sitting exams can need more than a week to recover from the strain: 80 percent of the students needed *eight days* to recover from poor sleep, distress, and other health complaints.[11]

TAKING A BREAK

While lifting weights in a gym, a break between sets helps you get your breath back. When it comes to mental work, the effects of a break are more heterogeneous. It can restore you to gear 2, or it can lower your gear right down to gear 1 so you can rest and recover.

The benefits of a break were shown to spectacular effect in a 2016 study on Danish schoolchildren taking standardized tests. The researchers found that test performance got worse with every hour of mental work, so the later in the day they took the tests, the lower their scores. A twenty- to thirty-minute break didn't just offset the deterioration, it enhanced their performance: had there been a break every hour, test scores would have *improved* as the day went on.[12]

Using our framework, the maximum time between two breaks should not be more than around ninety minutes (unless you are lost in creative thought). The break at the end of each ninety-minute work session should place you in a restful, defocused, "gentle curiosity" state of mind in gear 1 or near its boundary. Around fifteen to twenty-five minutes is a ballpark. If you don't want to lose momentum, limit your break to no more than ten minutes and take a longer break later, when you can.

The way you schedule breaks *within* a ninety-minute work session depends on the kind of work you are doing. Think of mental work as a sprint in this context. The more intense the work, the sooner your mental energy starts to falter and the shorter its duration ought to be before you break for a pause. Sometimes *how often* you take a break may be more important than *how long* your break lasts, so even the briefest of pauses can help. When taking a mentally intense exam, pausing for as little as five seconds every couple of minutes improves performance.[13] Taking a three- to ten-minute break every twenty minutes helps with passive, monotonous work.

It helps to think of a break in two ways: as a *charging station* and as a *rest station*. Within each ninety-minute session, a break can sometimes serve as a *charging station* to return you to gear 2 when you slide upwards or downwards. The break you take at the end of a work session acts as a *rest station*, rejuvenating your resources before you begin the session afterwards.

A break as a charging station

Sometimes your work might drive you into gear 1 because it is boring or monotonous. You are not tired as such, just understimulated. An example of this is long-distance night-time driving. At other times, your work might push your gear in the opposite direction and move you into gear 3 through overstimulation. Working in a noisy open-plan office with beeping alarms, or relentless multitasking, can have this effect. In both cases, simply stopping the work or escaping from the environment removes the upward or downward pull on your gear and immediately restores you to gear 2. If you take these "escape" breaks frequently, then you can keep resetting your gear to gear 2 as soon as it drifts upwards or downwards. In this case, a break acts like a *charging station*; it "charges" you back into gear 2.

You can incorporate activities into these breaks that act as a counterweight to the change in gear. For example, a short bout of exercise counteracts a lagging gear, and a breathing exercise offsets the tendency for your gear to creep upwards. In a lab experiment, paying continuous attention (in a way that is similar to long-distance driving) for half an hour led to a bigger decline in performance compared to interrupting the session halfway through with fifteen minutes of cycling.[14]

A Break as a Rest Station

When you want to rest and recuperate during a break, it can be helpful to first evaluate how you feel: do you feel tired in a relaxed or drained way, or do you feel tired and tense, or "wired"?

Tired and wired

Your muscles rest when you stop using them, but your mind doesn't necessarily rest when you stop working; it continues to work on your problems while you are on your break and grows even more exhausted. This is more likely to happen if you feel "wired" from your work and can't stop thinking about it. In such situations, you have to *actively* relax your mind to help it rest.

One way to stop your mind from ruminating on your work after you leave your office chair is by distracting it. You can accomplish this by doing something so absorbing—be it physical (a workout) or mental (a game like Tetris[15])—that there is no room in your head for anything else.

Another option is to "slow down" your mind, by doing something that lowers sympathetic activity and raises para-sympathetic activity to actively nudge you into a relaxed state, such as breathing exercises or yoga. One study found that doing a PMR (progressive muscle relaxation) session during their lunch break made call center workers less tired for the rest of the afternoon.[10] In another study, competing air rifle athletes recovered better after a break if they spent it doing breathing exercises, and this also improved their scores.[16]

Tired and not wired

If you are tired and *not* wired, your mind will behave more like tired muscle and fall into resting mode when you don't force it to work.

Taking a quiet, solitary break where you do nothing, speak to no one, and gently daydream or think of nothing, can help a tired mind recover. Walking slowly, idling, and reading a relaxing novel are also good options. You want to stay in or near gear 1, so it is wise to avoid emotional contagion from people, podcasts, and politics or anything else that raises your gear.

An even better approach is to take a nap.

The Best Kind of Break—a Nap

Earlier, I described the benefits of napping to prevent or mitigate the effects of the post-lunch dip, but napping can be mentally restorative at any time of the day. A fifteen-minute nap is usually enough for general alertness, but a slightly longer one is better for cognitive benefits.[17, 18] In one study on office workers who spent their days staring at a screen, interrupting a two-hour work session with a twenty-minute nap reduced fatigue and improved subsequent performance over taking a break without napping.[19]

Sleeping is a little like diving. Just as a diver must surface gradually or get the bends, your brain must awaken gradually if you have sunk too deep into the folds of slow-wave sleep or you get groggy, which makes it difficult to shift into gear 2 immediately upon waking. If you sleep without entering the slow-wave sleep phase, by contrast, you are less likely to feel groggy when you wake up. Although it is impossible to tell exactly when you will tumble into slow-wave sleep, a shorter nap reduces the chances of waking up groggy, though it is less rejuvenating than a longer nap.[20] Some researchers recommend napping for about twenty to thirty minutes to achieve that sweet spot, but there is considerable debate on this issue.[21,22]

One final point about napping is to try to nap as horizontally as possible. The flatter you lie when you nap, the better the quality of your sleep when you do.[23]

The Rhythms of the Mind

TAKING A BREAK

How often?	• Every 90 minutes **to rest.** • Every 2-10 minutes to reset and maintain the right level of alertness for work that is too underwhelming or too overwhelming.

How long?	• No more than 10 minutes if you don't want to lose momentum • Around 20 minutes **to rest** at the end of every 90 minute work session

What kind of break?	**To maintain alertness** *within* each work session:
	   • Choose activities that raise or lower your gear as needed. For example:
	      • Exercise to raise the gear
	      • Breathing exercise to lower your gear
	
	To rest *at the end* of each work session:
	   • If you are mentally tired and not wired:
	      • **Passive** relaxation
	         • Do nothing and think of nothing
	         • Do a "mindless" activity such as cleaning, tidying or walking in a green space
	
	   • If you are mentally tired & wired:
	      • **Active** mental detachment
	         • Do something immersive and gripping, like a crossword puzzle, or game on your phone
	      • **Active** physiological relaxation
	         • Stretch-release or contract-release activities such as yoga, PMR, or body scanning

The New Age of Knowledge Work

When management consultant Peter Drucker first coined the term "knowledge worker" back in 1959, most products falling off factory lines were things you could touch, such as cars, hair dryers, and toasters. They were made from raw materials that were also tangible, such as wood, metal, and plastic.

In today's knowledge economy, assembly lines have taken residence inside knowledge workers' brains, and they manufacture things you cannot touch: ideas, solutions, and insights. The raw material for these products is *information.*

You source information by continuously *learning,* and you then assemble the raw material into a product through *creativity* and *problem-solving.*

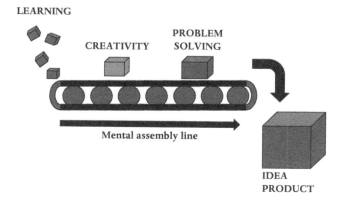

These three things—learning, creativity, and problem-solving—form the core of knowledge work. All are best performed in a gear 2 mental baseline, but you can improve your efficiency in each by playing with motivation, mental speed, and the way you pay attention.

7

FINDING YOUR INNER FLAME

"Don't ask what the world needs. Ask what makes you come alive and go do that, because what the world needs is more people who have come alive."

—HOWARD THURMAN[1]

As machines gnaw away at the routine and straightforward parts of knowledge work, they transform what was once a flat and predictable work landscape into a rugged terrain of hills and valleys.

Cycling along a hilly landscape demands a different set of skills from cycling on flat land. You have to pedal hard on a steep incline; the effort drains your stamina in a way that cycling on flat land does not. When you reach the top of a hill, you can stop pedaling and glide all the way back down with almost no effort at all. As you cross one hill after another, your effort surges and sinks in a rhythmic undulation.

In the evolving landscape of knowledge work, learning and problem-solving form "hills" that require a strong burst of power, and a rhythmic way of working perfectly complements this new terrain. As you ascend each hill to learn, solve, or create, you do so in gear 2 with brief forays into gear 3 for extra power. Once you arrive at the summit, you can rest your feet on the pedals for the descent in gear 1. With a rhythmic pattern as your base, you need three essential elements for successful hill navigation: *goals, motivation,* and *feedback.* A goal sets you on your journey. Curiosity and interest motivate your uphill climb. Feedback rolls you back down the hill, and the momentum of the whole process makes you want to do it all over again.

Many ecosystems for knowledge work are incompatible with this framework; work is not choreographed to be interesting or meaningful, productivity targets discourage rhythmicity of effort and output (and frown upon rest), and goals seem inflexible and remote. You can transform these ecosystems into formidable incubators for mental work by introducing a rhythmic way of working, embracing clear yet flexible goals, using intrinsic rewards to power your mental engine, and relentlessly chasing opportunities for incremental improvement.

SETTING GOALS

As knowledge work goes through a gradual metamorphosis, it starts to look increasingly like a kind of obstacle course. When you run on a dedicated track, or on a path in a city, your route feels certain and well-defined. An obstacle course, in contrast, is full of unpredictable hurdles and spontaneous detours. At every twist and turn, you have to clear your mind of everything else and formulate a plan of action. This plan is guided by a short-term goal that can sometimes run counter to your long-term aim. For instance, you may have to choose the slowest route down a muddy slope (to avoid slipping) even though your goal is to reach the finish line as quickly as possible. If you stick too rigidly to your overarching goal, you will not choose the best strategy for each challenge you encounter; but if you lose sight of that goal completely, you risk getting lost.

A fast-paced, rapidly changing knowledge work terrain is similarly strewn with unforeseen hurdles—be it a new skill you have to master, a creative breakthrough that needs to happen, or a problem that must be solved. These hurdles are best overcome in gear 2, and a goal helps to anchor your mind into a gear 2 state.[2] But the anchor must not be too rigid—you have to be prepared to change direction at a moment's notice every time a new hurdle appears.

To successfully navigate this kind of work landscape,

you need to be adept at shifting frictionlessly between (at times contradictory) short-term and long-term goals, without ever losing your overall sense of direction.

CULTIVATING MOTIVATION

When you encounter an obstacle in your work landscape, your mind needs a burst of engine power to overcome it. *Motivation* provides that power. It raises your gear and gives your mind a shot of energy so it can put in the effort required.

Motivation can come from within you or from the outside, and *where* it comes from affects *what* rewards you are drawn AmE toward.

When your motivation comes from *within* you, it tends to drive you AmE toward *intangible* goals — for example, the *desire* to learn something, get better at something, or overcome a challenge. You are not drawn to the material result of these things, but to the sensation of the process. This *intrinsic* motivation does something magical: it turns your mental engine into a self-driving car. You don't have to make it move — it moves all by itself, as if floating on the back of its own energy.[3] As Thurman says, it "makes you come alive."

When your motivation is engineered from the outside, or *extrinsic,* you tend to be drawn to *tangible* goals. Thousands of years ago, our primary goals were to acquire food and resources; today we tend to be more motivated by a paycheck, a job, or a marker of social status. Many of these goals are ones you have "learned" or been "trained" to carry. They tend to be tied to the *result* of a process, not the process itself.

This distinction is relevant in today's AI-driven knowledge work world because technology is disrupting the traditional relationship between a workplace and the tangible rewards it promises. In this new world, where learning new skills and solving unfamiliar problems can feel more like short term obstacles on the path to your larger goal, your motivation needs to be anchored in the process — in the intangible features of your work — rather than in its tangible outcome. This makes intrinsic motivation a powerful fuel-provider for knowledge work in the AI age.

In an episode of his podcast series *I Am* that featured a conversation with Steve Peters, the author of *The Chimp Paradox,* English rugby champion Jonny Wilkinson shared a personal account of how the two kinds of motivation, intrinsic and extrinsic, skewed his perception in radically different ways during critical moments of play. Wilkinson recalled how every rugby game would envelop him in an almost exhilarating flow of effortless play, but he snapped out of this zone the moment the referee blew the whistle signaling the need to kick a penalty. Dreadful "what if" thoughts would flood into his mind as he suddenly became aware of the immensity of the moment and of the consequences of failure. Suddenly he was no longer driven by the intangible, almost childlike pleasure of playing, but by the tangible *result* of his performance—the need to prove his worth to the world and not let his country down. The referee's whistle switched off intrinsic motivation and turned on extrinsic motivation in a split second.

Motivation Through Outcome

Most workplaces rely on *extrinsic* forces to motivate people because external motivators are easy to manipulate. Financial incentives, material rewards, or a promotion are some examples. The desire to *avoid* something tangible—whether missing a deadline, losing your job, or forfeiting a promotion—can act as an external motivator, too. In the era of assembly lines, the temptation of good wages, the lure of social status, or the threat of losing one's job in the factory served to keep exhausted workers energized through their desperate boredom. In the knowledge work era, businesses and corporations famously use bonuses, large salaries, and promotions to coax weary workers to carry on.

When you're relying on extrinsic motivation, everything rests on the outcome. Because outcomes carry an element of uncertainty, extrinsic motivation necessarily carries some psychological tension. People who perform best under pressure usually thrive under external motivation: the thrill of a prize or the danger of a punishment can be a potent motivating force for them. If your gear is "springy" and you are sensitive to uncertainty, then external motivators can haul you straight into gear 3. This explains why a student who comfortably solves a math equation in the classroom on a quiet Friday afternoon (in gear 2) is stymied by the same equation during a high-stakes exam (gear 3).

Turning threats into challenges

An extrinsic motivator can take the form of a challenge or a threat. Both raise your gear, but a threat can raise your gear more potently, by exciting brain networks that compute emotion. Imagine you just started a new job, and your CEO announces she will be spending time with your small team. You could see this as a challenge to earn her respect, or you could see it as a threat and worry about making a bad impression.

A threat makes you *play to avoid a loss*. A challenge makes you *play to win*. A magic ingredient can turn the former into the latter: *controllability*. The moment you find something you can control, a threat becomes conquerable and downgrades into a challenge. If you had been catapulted to gear 3 by a perceived threat, you are drawn back into gear 2. This also works in the reverse direction. If you are confidently tackling a challenge and suddenly lose control, the challenge starts to feel like a threat. What feels like a threat to one person can feel like a challenge to another because two people can perceive different levels of control in the same situation.

In 2008, sport researchers Geir Jordet and Esther Hartman looked at "all penalty shoot-outs ever held in the World Cup, the European Championships, and the UEFA Champions League" and made a striking observation: footballers could have scored almost 30 percent more goals if they had been motivated by the thought of a win rather than the fear of loss.[4] When missing the goal meant losing the game, only 62 percent of kicks were successful. But when scoring a goal meant winning the game, that number rose to 92 percent of kicks. The threat of loss motivates you through fear, which drives you into gear 3, where you lose focus and fine control.

Athletes are famously coached to "play to win," but this mindset becomes progressively difficult to maintain as an athlete accumulates a series of wins and can't help playing to avoid "losing" their reputation. (Consider how, when a Wimbledon champion returns to play the following year, they are officially called the *defending* champion.") The same principle applies to professional success. In the early stages of our careers, we want to advance and conquer new territory, but as soon as we "make it" to a top position, we switch to a defensive mindset and want to protect the position we have.

Any workplace that encourages a "blame" culture forces workers to adopt a loss avoidance mindset. People "try not to fail" rather than "try to win," and the fear of punishment can nudge many into a perpetual gear 3 state of mind. Since creative thinking is stifled in gear 3, ecosystems that punish failed ideas or ventures — whether financially or psychologically — destroy the spirit of innovation and entrepreneurship, two talents that are essential for success in this new knowledge age landscape.

The two sides of competition

Competition is inherently "uncontrollable" because you cannot control your competitors — but you can tune its threat level down by increasing your *perception* of control. When competition happens in a culture of fairness, safety, and transparency, it feels motivating, but not so in the presence of opportunism, plagiarism, and dishonesty. As RK, who manages software innovation in a large corporation, put it, "competition can spawn new, often breakthrough ideas, as long as everyone works by the same rules, the system is open and transparent, credit is always given where it is due, and plagiarism is not tolerated." He has discovered that protecting people's self-worth is key to cultivating such a culture. "When people feel valued and the effort of idea generation is rewarded as well as the idea itself, they feel more in control." This control allows you to tap into the intrinsic rewards of the work, turning competition into a motivator instead of a threat.

In his previous position at another company, RK experienced a very different culture. There, the individual did not matter as much as the ideas that were generated, and dishonesty was widespread. RK left that organization because it was riddled with high rates of burnout and employee attrition.

The perception of threat invites a defensive, often emotional reaction and pushes you into gear 3, which paralyzes your capacity to imagine, daydream, and let your mind wander. Competition therefore needs careful calibration in a mental factory whose main product is idea generation — because *you cannot create while you defend*.

Intrinsic motivation

It can feel impossibly difficult to muster interest in the work you have been assigned to do on a cold Monday morning. But if you manage to find a spark somewhere, that spark can make even the most mundane work feel lighter.[5] That spark makes the work "come alive." Whatever you are doing, you will *want* to do it, you will *enjoy* doing it, you will want to *keep* doing it, and you will be able to swerve easily into the right state of mind.

Finding Your Inner Flame

When giving a talk at AT&T Bell Laboratories in New Jersey in 1952,[6] Claude Elwood Shannon — whose landmark research paper on information communication sowed the seed for the information age — pointed to intrinsic motivation as the defining trait shared by the small number of people who produce "the greatest proportion of the important ideas." Whereas training, experience, and talent are important, Shannon argued, they are not enough to produce genius breakthroughs. The formula behind Einstein and Newton, he surmised, is "some kind of a drive, some kind of a desire to find out the answer, a desire to find out what makes things tick."

It is easy to motivate someone to do something with a cash incentive, but the inner desire to do something for its own sake is so intangible and elusive that it is almost impossible to "turn on" from the outside, on command.

Learning Progress

One intriguing strategy, however, does seem to reliably "switch on" intrinsic motivation: turning the process of work into a journey of progressive improvement, or what researchers Pierre-Yves Oudeyer and Frédéric Kaplan term *Learning Progress*.[7,8]

Change is a defining characteristic of the world we live in. Where there is change, there is uncertainty. But uncertainty hides danger, so it is existentially wise for us to want to reduce it. The way we do this is by *learning*. Whether you learn to acquire knowledge or to build competence and skills, you make the changing world a little more controllable with any progress you make.

This is where the *Learning Progress* mechanism comes in. The mechanism rests on the idea that we are wired to continually strive to improve ourselves and master the unknown around us. Whether you learn by amassing knowledge or building skills, the sensation of making progress feels deeply satisfying and energizing. This fuels an intrinsic motivation that drives you forward.

You can think of the difference between learning and Learning Progress as similar to the difference between doing and *improving*. Learning a set of isolated facts or skills will not evoke a sense of progress—unless doing them *progressively* narrows the distance to some kind of a goal. You need to be consistently aware of making progress through some form of feedback.

Historian Anton Howes, the author of *Arts and Minds: How the Royal Society of Arts Changed a Nation,*[9] believes that a reason for the boom in innovation in Britain in the eighteenth century was that people "adopted an improving mentality." The Learning Progress mechanism would suggest that this desire to improve—whether by mastering uncertainty or by upgrading skills—is woven tightly into the fabric of the human condition but remains bridled and camouflaged by the toil of everyday life.

Much of Oudeyer and Kaplan's research is inspired by the way babies and young children are moved to learn through wonder and curiosity, without any thought of a goal or reward. If you take a three-month-old baby and gently tether one of her feet to a mobile toy, an intriguing transition takes place. At first the baby kicks with random abandon. Then, as she notices that each kick makes the mobile toy move, she is jolted by an *a-ha!* moment of insight as she realizes, *"I can make the world move!"* This realization flips some kind of a switch and sows the seed of agency in the baby's mind. From here onwards, the baby stops being random and behaves with purpose, confident that a continuum exists between herself and the world such that an action always produces a result.[10]

This bond between an action and a result doesn't just satisfy the baby: it instils an inner desire to perform the action again. With every new kick she learns a new detail, whether the precise force needed to make the toy move a certain distance or the direction to kick in for better entertainment. The baby also learns more about her own foot: perhaps the sensation of the kick, or how to wriggle her toes in just the right way. The feedback confirms that there is Learning Progress. The extraordinary thing about this whole effortful endeavour is that it happens without any external bribe or compulsion. The baby is left to her own devices and burrows herself a rabbit hole of exploration along the foot-toy continuum. The process isn't a chore for the baby; on the contrary, it seems deeply pleasurable.

The *progress* part of the Learning Progress mechanism is key because it prevents you from gorging on information indiscriminately or becoming overwhelmed through pointless endeavours. If an activity is too difficult, progress will be impossible. If it is too easy, there will be no opportunity for progress. The optimal level of difficulty lies somewhere in the middle.[11] Learning Progress also keeps your focus on the next step instead of on the finish line, which makes your journey less daunting (you're not constantly reminded of how far you have to climb) and efficient (you carefully consider every step before taking it).

Learning Progress may help explain why different people feel the joy of intrinsic motivation while doing slightly different things: your progress will be fastest if you already have a base to start from, or if you have a natural flair for what you are doing. Learning Progress—and hence intrinsic motivation—is therefore most likely to emerge from activities that complement your unique portfolio of skills and talents.

The Learning Progress mechanism points to two "entry" portals to prime your mind for mental work.

Imagine you're sitting at your computer first thing on a groggy Wednesday morning, looking at a cumbersome report your boss has emailed you. You need some kind of hook to latch your attention onto, and two strategies can help:

1. Find Learning Progress

   Cast your eye over the whole report until you come upon something, no matter how trivial, that you find interesting. Begin the report there. Use the momentum of your progress to find a second crumb of interest elsewhere in the report, and use that progress increment to find a third, and so on. Instead of going through the report sequentially, chip away at it in no particular order using only Learning Progress as a guide. The process is similar to filling in a crossword puzzle: you tackle the clues in the order that you can solve them, and every word you complete makes the whole thing incrementally easier and more interesting.

2. Manufacture Learning Progress

   If nothing in the report sparks your interest, then actively manufacture a sense of Learning Progress. Segment the report into a series of actions that you can measurably progress along. Add a feedback signal at the end of each action, such as a box you must tick. This progress-feedback volley inspires a sense of Learning Progress.

The 80 percent heuristic

The difficulty level of your work influences how much Learning Progress you make and hence its potential for sparking intrinsic motivation.

In 2019, U.S.-based researchers trained a neural network to mimic the way the human brain learns new information.[12] The learning exercise resembled the way a radiologist might learn to identify a fracture on an X-ray, with a binary "yes" or "no" answer. With each X-ray the radiologist gets wrong comes the opportunity to learn: "A fracture can look like this, too!" In the neural-network-based paradigm these researchers tested, learning happened fastest when the agents were getting about 15 percent of the answers wrong.

The number of incorrect answers gives you a sense of the difficulty level of the test. If you are correct 50 percent of the time, the test is so difficult that you are doing no better than chance — and not learning anything from the process. If you are always 100 percent correct, the test is too easy and there is nothing for you to learn. At a 15 percent error rate, the test is neither too difficult nor too easy. This finding implies that Learning Progress is optimal at *moderate* levels of difficulty.

A 15 to 20 percent difficulty level makes intuitive sense as a heuristic for rapid learning. If you know nothing about a subject, then learning it feels frustrating. But if you already know a lot, there isn't much motivation to learn more. This heuristic is rooted in the neural dynamics playing at the level of your gear network. When something is too easy and predictable, you are bored into gear 1; too great a challenge takes you into gear 3. In the middle Goldilocks zone, you are in gear 2.[13]

An approximate way to arrive at this Goldilocks zone is by matching the difficulty level to your skills so they are slightly stretched. When you stretch your skills, *you learn*. When you do it again and again, you make (learning) *progress*. Your work should be difficult enough to challenge you, but easy enough for you to perform the work by stretching your skills.

Pleasure **after** *effort*

When our ancestors went hunting thousands of years ago, the quality of prey they acquired would have been directly proportional to the effort they applied. Large animals had the most meat, but they tended to be the hardest to trap, whereas smaller animals were easier to trap but promised fewer calories. For the most part, this would have established a direct bridge between the effort you put in and your enjoyment of the fruits of your labor: the bigger the effort, the greater the reward.

The sacred marriage between effort and reward, instilled into the human psyche over millennia, became strained when industrialization contorted their relationship in both time and space. Industrialization has added so many layers of complexity and convolution between effort and output that a single keystroke could have market-altering implications across half the globe; at the same time, hours and hours of meticulous coding could result in one trivial product improvement that users barely notice.

Yet this impression created in the time of our long-gone ancestors lives on in our minds: while effort exerted doesn't necessarily impact the size of the tangible reward you receive (the athletes who train the hardest don't always win), it does affect how satisfied you feel afterward. Recent laboratory experiments show that your prize is more likely to bring you joy if you have worked hard for it.[14]

Pleasure *through* effort

We carry a curious bug in our mental software. If you *consistently* reward intense effort, a strange short circuit emerges.

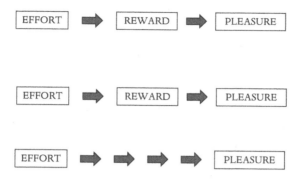

If you reward people for their effort over and over, the effort itself turns into a kind of reward. Putting in effort generates the same joy that a reward would generate; the process of trying hard becomes pleasurable in itself. A European research group recently demonstrated that when people were rewarded after they put in mental effort, they were more likely to choose mentally effortful work in the future, even if the effort did not lead to a reward.[15]

Once you're able to summon this kind of joy simply by putting in effort, the effort of engaging with your work generates its own intrinsic motivation, and the strain melts away.

It may not matter where your brain learns to associate effort with pleasure for you to experience the short circuit. Investing in effortful, intrinsically rewarding activities outside of work can compensate for a distorted effort-reward balance at work. A veteran trader at an investment bank once told me that the members of his team who remained the most driven during a disappointing day were those who frequently engaged in intense challenges and extreme sport — ultramarathons, adventures, mountain climbing — during their time off. They put effort in, even if it was pointless, because they enjoyed the effort for its own sake.

If you take the ingredients for intrinsic motivation — consistent, incremental improvement; a perfect match between the level of difficulty and your skills so that your skills are partly stretched; and the pleasure of achievement after the effort of trying — and blend them in the right ratio, you can create a state of mind where intrinsic motivation reaches a hiatus: *flow*.

FEATURES OF A HIGH STRESS WORKPLACE THAT
RAISE AND LOWER YOUR GEAR

LOWER RAISE

WORK CONTENT	LOWER	RAISE
	• Well-defined goals	• Nebulous objectives
	• Clear instructions	• Ambiguous instructions
	• Opportunity for learning progress	• Lack of autonomy
	• Agency & autonomy	• Effort and reward mismatch
	• Effort and reward matching	• Punishment based incentives
	• In-person communication of complex issues	• Reward based incentives
	• Low frequency/volume of incoming information	• Digital communication of complex issues
	• Working to "human time"	• High frequency/volume of incoming information
	• Longer, less frequent deadlines	• Working to "technological time"
	• One-task work sessions	• Tight deadlines and many "urgent" demands
	• Low work volume	• Frequently changing tasks
	• Lower difficulty level	• High work volume
		• Skill-task difficulty mismatch

WORK ENVIRONMENT	LOWER	RAISE
	• Safety & cohesion	• Competition & hostility
	• Collaborative culture	• Blame culture
	• Order & predictability	• Chaos & unpredictability
	• Transparent communication	• Opaque communication
	• Low frequency/volume speech	• High frequency/volume speech
	• Warm soft light	• Cool, harsh lighting
	• Silence, or white noise	• Distracting noise

MENTAL/PHYSICAL	LOWER	RAISE
	• Boredom	• Stress, worry, rumination
	• Napping	• Caffeine
	• Breathing exercises	• Exercise
	• Stretching	• Working at full power despite fatigue
	• Breaks	

8

FINDING YOUR FLOW

"In the middle of this crazy action, the ball's flying, and it's a hundred miles an hour, and yet within it, it seems like there's a serenity within me . . . just navigating the route almost effortlessly."[1]

—JONNY WILKINSON

Wilkinson is talking about one of the most puzzling phenomena of mental performance: a state of mind in which you become so immersed in what you are doing that the world around you all but disappears from your consciousness, you barely notice the passage of time, and your work feels effortless. About half a century ago, Hungarian psychologist Mihaly Csikszentmihalyi noticed that when people reach this state of mind, two seemingly contradictory phenomena emerge at the same time:

- You put in *less* effort because the work suddenly feels lighter.
- Despite putting in less effort, you perform better.

It is almost as if, instead of depleting your energy, your work *replenishes* your energy when you are in this state, and your performance reaches a peak. Csikszentmihalyi called this state of mind "Flow."

For Flow to emerge, there needs to be a challenge you overcome by *stretching* your existing skills. You also need immediate and *clear* feedback that you have overcome the challenge, as receipt for your effort. The pleasure of overcoming the challenge motivates you to want to do it all again.

Your work is most likely to induce Flow if it is structured as a steeplechase—a series of challenges or steps where each challenge or step is followed by some kind of feedback signal that confirms you have been successful. The signal acts like a "prize" for your efforts. The tension of each challenge or step pushes your gear upward, and the signal at the end pulls it back down. The pleasure you feel after conquering a challenge makes you want to do it again. As you proceed along the steeplechase of effort and feedback, your mental state undulates in a rhythmic cycle of tension and pleasure.

In the world of mental work, Flow is a kind of superpower. Keeping your mind running at maximal efficiency for long periods feels exhausting because you have to work hard at pushing and pulling your gear up and down at regular intervals to stay in an optimal zone. When you're in a state of Flow, however, this push and pull seems to occur almost effortlessly. How can work that usually requires great effort suddenly be done *better* with *less* effort? What makes such work become sustainable for hours on end?

In 2009, René Weber, professor of media neuroscience at the University of California, Santa Barbara, and colleagues from UCSB and Michigan State University came up with a theory rooted in an extraordinary feature of the brain: its rhythmicity.[2] Many researchers view the brain as a constellation of pendulums. There are large pendulums and small pendulums, ones that move fast and others that swing slowly. Memory, attention, and learning depend on pendulums rippling across brain networks.[2]

Pendulums have a remarkable property: if you swing two pendulums from a common base side by side, they *synchronize*. Think about what happens when audiences clap for an encore at the end of a show or concert. At first the clapping is jagged and chaotic as everyone claps to their own rhythm. Then suddenly, out of nowhere, everyone starts clapping in synchrony. No one actively tries to do this, or consciously plans it; it happens spontaneously. This sets the scene for Weber's theory.

As you focus your efforts to overcome a challenge, the networks in your brain that are responsible for **attention** fire vigorously. When you receive feedback that you have overcome the challenge, your brain's **reward** networks do so, too. The two networks normally fire at their own rhythms, but as you go through a rhythmic steeplechase of effort and feedback, they fire in dialogue, and just like the clapping hands, they *spontaneously synchronize.* This is when you enter Flow. The synchronization isn't gradual: it happens instantly, just like the sea of hands clapping at the end of a concert.

When things synchronize, they acquire a peculiar feature: they become more efficient. Think back to the audience clapping at the end of a concert. When the audience synchronizes its clapping, the volume of clapping grows louder because the claps are falling at the same time. The *same* effort of clapping produces a *bigger* volume of noise.

By a similar analogy, when the attention and reward networks in the brain synchronize, they become more efficient. This is why you need less *effort* to do the same amount of mental work in Flow. One way to picture this is to imagine a circus juggler with spinning plates. It takes a lot of effort for the juggler to balance a plate on his toe, on his nose, and in each hand. But if the juggler can get all the plates spinning in synchrony, balancing them suddenly becomes much easier.

Weber's theory suggests that the boost of energy we experience when in the state of Flow is rooted in this efficiency. To nudge your brain into the Flow state, you need to enter gear 2, then establish a volley of push and pull forces that turns into a self-sustaining spin.

APPLYING FLOW TO KNOWLEDGE WORK

It is tempting to think of Flow as some kind of evolutionarily preserved switch that transforms the brain into a self-propelled machine that derives joy from the act of upgrading itself and improving its existence. Given how the Flow state makes the act of stretching yourself beyond the limits of your comfort zone feel inherently pleasurable, it is an invaluable tool for success in a knowledge workspace disrupted by rapid change.

Five core conditions are necessary for any work to have the potential for Flow:

1. **Challenge:** You need to be in pursuit of something challenging.
2. **Clear goals:** You have to know how to go about the challenge.

3. **Skill match:** Your skills should be stretched, but not overwhelmed.

4. **Immediate, clear feedback:** You should receive clear and regular, real-time feedback.

5. **Motivation:** You must want to overcome the challenge over and over again.

The game Tetris is known for inducing a Flow-like state. Tetris is a puzzle-style video game in which you have to fit falling tetrominoes into congruent vacant spaces as quickly as possible. When a tetromino appears on your screen, it poses a challenge. You feel relieved as soon as you fit it into place, and the pleasure of the experience makes you want to do it again. You play at a level that stretches you, and the more you improve, the harder it gets. The shapes fall faster and faster until the effort-and-pleasure dialogue spins into a volley. You soon lose yourself in the game and it becomes effortless.

The steeplechase of effort and feedback riding on challenge and its resolution can run along different time frames. If you are creating a new design for a product, for example, the challenge persists until you have come up with the final design, but it temporarily abates every time you have an idea. In sports, the challenge rests in ultimately winning the game, but you experience temporary resolution when you make the move you had rehearsed, when you receive haptic feedback, or when you score.

Similar patterns can be found across a variety of knowledge work landscapes. Here are some examples:

Software engineering

Coding is famously Flow-inducing. The code to be written can be divided into segments, with each segment becoming a discrete challenge. Running the code successfully at the end of each segment and then moving on to the next segment maintains the challenge-resolution volley.

Creativity: Art, writing, music

There are endless anecdotes describing how artists, writers, and composers are often tormented by an unresolved concept that besets their minds until they resolve it into tangible form. The challenge temporarily resolves into relief as a chapter or line of music takes shape, creating an ongoing challenge-resolution volley until the work is complete.

Design

A designer of any variety carries an unrealized idea in their head. The ongoing challenge is to corporealize it. Every idea, every design decision momentarily resolves the challenge and feels rewarding; then the next challenge surfaces and the cycle takes place all over again.

Learning

Online learning courses make use of these dynamics through the precise engineering of Learning Progress by tailoring difficulty to test performance after each module. Every module in a series introduces new material at a pace that keeps test performance at around 80 percent. If test scores sink lower, the subsequent module is easier, and if scores improve, it grows more difficult. This maintains an optimal level of challenge, and the tests provide feedback.

Gamification of the workplace

Workplaces are increasingly using strategies from the gaming industry to make mundane work less boring and inspire workers to enjoy what they do, in an effort to improve productivity. Many of these strategies cluster around the main principles of challenge and feedback and hence revolve around the concepts of intrinsic motivation. If you are quietly stacking shelves in a warehouse, your work will immediately feel more interesting if you compete with a colleague to see who stacks shelves faster. A "progress tracker" that beeps and earns you a badge every time you have moved a certain number of items will also lessen the drudgery. The first approach provides a challenge; the second offers you feedback.

If you combine the two, the tedium of shelf-stacking becomes a two-player game. With some adroit fine-tuning, the game could even put you in Flow and make you *want* to stack shelves for the fun of it.

In his book *Actionable Gamification: Beyond Points, Badges, and Leaderboards,*[3] author Yu-kai Chou emphasizes that badges must be awarded at the right moment — immediately after a challenge is overcome — to have the most impact. This underscores the importance of the challenge-feedback volley in Flow, and of the effort and reward relationship fundamental to intrinsic motivation. As well as competitions and badges, some workplaces use dashboards (with real-time analytics) and to-do lists (with points earned for every box ticked) to maintain the momentum of challenge and feedback.

INSPIRING INTRINSIC MOTIVATION IN THE WORKPLACE

Although not every aspect of knowledge work, whether trawling through lines and lines of code or solving insuperable problems or reading a report in legalese, can be gamified, even the dullest task can be engineered to bear some amount of challenge and pleasure.

A workplace that gives you agency — where you have autonomy over the tasks you do, and the order and pace at which you do them, and where you can consistently and incrementally improve in some domain — is more likely to hold opportunities for intrinsic motivation. Material incentives that pull at your strings from the outside antagonize intrinsic motivation. If you are doing something to win status, for a financial incentive, or for some other tangible reward, your eyes stay on that prize and you neglect the journey from which the inner pleasure derives. Intrinsic pleasure arises in the *sequence,* not in the *consequence,* of your actions.

Here are six ways to curate an ecosystem where people can enjoy work for its own sake:

1. Define everyone's role clearly, so the team knows whom to credit when something gets done.
2. Give everyone a clear road map so they know what they are expected to do and how to go about doing it. This lets everyone know where the effort-reward ratio will be maximal.
3. Identify everyone's personal passions, skills, and goals; delegate tasks that align with those whenever possible.
4. Acknowledge the *process* of effort, regardless of its consequences, and reward it in some way, even with a rain check gesture.

5. Nurture a culture of unyielding fairness. When rewards (including praise and recognition) are distributed unfairly, workers stop expecting that their effort will lead to reward, which severs the challenge-resolution volley.

6. Monitor fatigue, boredom, and stress meticulously. These push you out of the mental state synonymous with intrinsic motivation (gear 2) and make flow impossible to attain. If these are the result of a skill-challenge mismatch, then correcting the mismatch will optimize conditions for intrinsic motivation.

A common theme runs through all the routes that lead to intrinsic motivation: expansion into the unknown. Flow emerges when stretching your skills; learning progress, by its definition, implies learning what you don't yet know.

The tension you feel when you sense uncertainty, when you stretch yourself to meet a challenge and mobilize your resources to overcome it, reflects the surge in norepinephrine released by your blue dot network. That norepinephrine diffuses across the nerve endings in your brain, priming them to make new connections. Your brain becomes more malleable and moldable; it is easier to shake off old rules and learn new ones. This process may partly explain why people sometimes describe a sense of exhilaration, of "feeling alive," during moments of intense challenge. In a sense, the tension comes from colliding with the unknown, and the aliveness comes from the thrill of stepping into it.

We may have evolved to "feel alive" when expanding into the unknown because the drive to extend our territory of "known existence" *has kept us alive.* The incessant drive to keep updating and augmenting our knowledge would have helped us adapt in the event of sudden or catastrophic change; it saved our ancestors' lives—and gave us our own.

9

LEARNING AT THE PACE OF CHANGE

"The very same process of automation that causes a withdrawal of the present workforce from industry causes learning itself to become the principal kind of production and consumption."

—MARSHALL MCLUHAN[1]

Writing in *California Management Review* in 1999, some forty years after introducing the concept of knowledge work to the world, Peter Drucker listed the factors he deemed most essential for knowledge worker productivity: *continuing innovation* was one; *continuous learning* was another.[2]

The world around you is constantly changing, sometimes by a small amount, at other times by far more. Every change, little or large, introduces a gap in your knowledge, which your brain quickly fills by learning. We have, however, entered an unprecedented phase in human history, when change is outpacing the speed at which the brain is able to learn.

In a 2017 Aspen Institute lecture, Thomas Friedman describes how Eric "Astro" Teller, the CEO of Google X (now X), once explained to him that the rate of technological progress has overtaken the rate of human adaptability. He drew the relationship between the two on a graph. A smooth line with a small gradient represented human adaptability. A second line began beneath the first, curved upward to meet it, then shot beyond, exponentially. This second curve illustrated technological advancement.

The "Teller Curve"

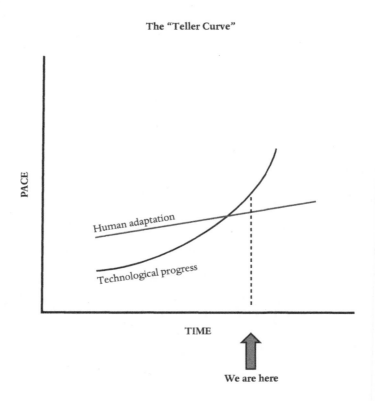

PACE

Human adaptation

Technological progress

TIME

We are here

According to Teller, we have just passed that intersection. As the technology curve races ahead, new situations, products, ideas, information, and problems land in front of us at dizzying speed.

In the past, a block of knowledge acquired through school and a college education would last us a lifetime. Today it risks growing obsolete within mere years. I have seen this in my own career as an eye surgeon. When I passed my board exams, a scanning technique called optical coherence tomography (OCT) was not even part of my syllabus. A few years later, fluency in OCT became as essential for an eye surgeon as knowing how to read a chest X-ray is for a lung specialist.

To keep up with the soaring technology curve, we will need to continuously upgrade skills and expand knowledge across our lifetime. The divide between success and failure, Friedman says, will no longer be decided by the trajectory of formal education, but by the ability to keep learning long after leaving college. An evolutionary pressure will select those who can learn well from those who can't, and "the motivational divide will be more important than ever."[3]

Learning in the age of AI-assisted knowledge will resemble learning for an exam with a constantly changing syllabus. Although you will have a goal at every point, that goal will relentlessly change. Sometimes the goal will be to learn a new skill; sometimes it will be to learn new information; and at other times it will be to learn and retain complex concepts and ideas. Adjusting your gears for each type of learning will help you do all three efficiently.

LEARNING SKILLS

One of the more curious features of our current phase of the knowledge age is that knowledge workers are not the only ones who must learn—the tools learn, too! Many kinds of AI technologies improve themselves through continuous learning, and if their learning rate outpaces yours, then they change faster than you can learn to use them. Large language models are one example; if a new one comes along before you have mastered the previous one, this leaves you with little choice but to discard traditional ways of learning and instead get comfortable with plunging right in and learning on the job.

Earlier I mentioned how the uncomfortable tension you feel when you encounter the unknown is the sensation of your blue dot network firing faster and releasing more norepinephrine. But remember, this surge in norepinephrine *catalyzes learning*, so the discomfort you feel when you must navigate an unfamiliar challenge or unexpected situation is a signal that your brain is optimizing its capacity to learn.

If you face so much uncertainty that this discomfort becomes excessive, your brain will tip over into gear 3, where you cannot reliably learn new facts, ideas, or concepts. If you can hover in the high-energy gear 2 zone *without toppling over into gear 3,* your ability to learn on the job will increase dramatically.

When FZ, now a senior tech executive, began as a junior software engineer at a major tech company, he found everything to be frighteningly new and unfamiliar: the system, the libraries, the design paradigms, and so on. Afraid to modify the codebase before figuring out exactly how it worked, he made very slow progress. After three weeks, his irate manager chided him for his lack of contribution. In desperation, FZ opted for Plan B: he would plunge in despite his limited knowledge. That afternoon he dived into the code and narrowed in on a section that was least unfamiliar, where he made a very small change. The code didn't break. Feeling marginally braver, he did it again, this time casting his net slightly wider into the unknown. Yet again, nothing broke. FZ went on in this way, gathering speed with each new challenge. Looking back, he told me he became proficient faster than he would have by just reading the documents and studying the code before diving in.

GK, a manager at a global investment bank, told me a similar story. When he first became a manager, he was assigned a project that fell entirely beyond his sphere of expertise. What's more, he had a new team that hadn't yet learned to trust his leadership. Having no choice but to plunge in, he first looked for aspects of the project that were common to the ones he was already familiar with — data gathering, people management, and communication—and tackled each of these, a small step at a time. This sense of progress gave him the confidence to venture further into unknown areas, and his foundation of knowledge gradually expanded. GK told me, "It is much better to start immediately with where your skills maximally overlap, and grow in the other areas while getting a sense of achievement along the way, even if it isn't as efficient."

In the age of AI, when your tools learn faster than you, learning how to dive into the unknown—and feel comfortable in the process—will become an invaluable skill. You can transfer the skill to any situation and any kind of work—and if you can do it without fear, it will give you a considerable edge.

LEARNING INFORMATION

The mechanics of learning involves capturing the information you need and holding it on a "temporary" memory shelf until it is filed away into long-term memory.[4,5]

Managing your attention is critical when you learn. You can't take in information that you are not paying attention to, so if you are too distracted while you learn, the information won't enter your head in the first place. This is why learning simple facts, concepts, and information is easiest in gear 2.

Once you have the information on your temporary shelf, you must take care not to get distracted by new, more exciting facts or ideas that can evict the original bit of information. Briefly turning your attention inward *immediately after* learning something can mitigate this "eviction risk" and help you remember what you have learned.[6] This might explain why walking, which prevents your attention from sticking to anything around you, can help when learning vocabulary or other factual material.[7]

Slightly later, once you are out of the window of "eviction risk", directing your attention to cues that shift your gear upward can help to cement your new knowledge.[8,9] This may partly work by summoning more norepinephrine, which augments learning. When some people had to learn a list of words for an experiment, those who watched a stimulating video afterward remembered more words the next day than those who did not.[10] In another, "real world" experiment, students who watched an exciting video clip after sitting through a psychology lecture did better than their peers when tested on the material two weeks later.[11]

The role of exercise

A French research team recently showed that the brain may not be able to rely entirely on sugar (or glucose) as fuel when it takes on a heavy cognitive load.[12] To compute high volumes of information, it needs a second type of fuel: *lactate*. One potential path (of many) through which exercise supports mental performance is by temporarily increasing levels of lactate in the body, which then crosses into the brain.[13]

Exercise nurtures your brain's ability to learn and store new information in a plethora of ways, but many of these routes are nonlinear, dispersed in time, and variably activated by different kinds of exercise. Consequently, it can be difficult to predict exactly what effects a single session of exercise will have on a specific instance of learning.

For example, when some people were asked to memorize the locations of different pictures on a screen in one study, those who exercised four hours after viewing the pictures were better at remembering the information two days later than those who did not exercise or who exercised immediately afterward.[14] In another study, a couple of three-minute sprints, done fifteen minutes before learning vocabulary, improved the rate of learning by 20 percent; curiously, running less intensely for forty minutes did *not*.[15]

Exercise enhances learning by raising your gear and inviting a rush of norepinephrine. You can harness this effect to enter a high-energy gear 2 state where you learn the fastest. Exercising too hard, however, can leave you feeling so tired that you slide into gear 1 afterward, where learning is suboptimal. The net effects of exercise will depend on how tiring it feels, your baseline energy that day, and the complexity of what you are learning.

If you're exercising *during* learning, the exercise will help only if it is light, such as walking, in part by keeping you in gear 2. In one experiment, gentle cycling helped people perform better only if they were weak performers to begin with.[16]

LEARNING SIMPLE FACTS

15 minutes before	• A short, sharp, *burst* of exercise: e.g. 2 x 3-minute sprints separated by a 2-minute rest break
Immediately afterward	• Don't let your memory be flooded by any new or exciting content • Take a walk • *Avoid* moderate or intense exercise
10-30 minutes later	• Listen to fast-paced music • Watch an interesting video clip
4 hours later	• Exercise at moderate intensity for thirty minutes

LEARNING COMPLEX CONCEPTS AND IDEAS

If learning simple information is like filing a book on a shelf, learning complex concepts is akin to filing a book on a specific spot in a huge library whose contents and configuration are always changing. Your mind must fit the new ideas into the dynamic web of knowledge you already possess, and this calls for agile maneuvering across the spectrum of gear 2 states. There will be moments of deep focus and times when you have to loosen your attention and pursue branches of thought far into the periphery, wandering in a gentle way or joining dots with excited curiosity. At other times, you will beam your focus inward, into your mind, as you rearrange your mental library. Going for a walk is one way to do this, and there are many anecdotes of people who spend hours struggling to comprehend a concept and then get sudden clarity during a long stroll.

FLASHBULB MEMORIES

If you think back to the last time you felt physically shaken by something *positive or negative* when you were given terrible news or were elated by an event, you will remember every detail of the scene that played out in a way you wouldn't have if the context had been mundane. This kind of "flashbulb" memory—where you remember moments almost viscerally and recall details of scenes, sensations, and scents mired in emotion—is forged in gear 3. You cannot learn anything that requires you to think and focus in this mental state; for that to happen, you need to be in gear 2. In a "flashbulb" memory of your final math exam at school, for example, you may not remember the questions on the exam paper, but you will forever remember the scent in the hall where you sat the exam.

WHEN EMOTIONS LEND A HAND

Emotions can help the learning process, partly by gilding what you are trying to learn so it stands out in your mind. A New York–based research group recently showed how attaching an emotional meaning to otherwise neutral sentences enhanced the ability to learn the sentences verbatim.[17] But remember, if you are learning information or concepts, the emotions must not push you out of gear 2 *while* you learn. Only flashbulb memories are possible when you cannot think straight on account of emotional turmoil.

10

THE CREATIVE MIND

"Big ideas come from the unconscious. But your unconscious has to be well-informed, or your idea will be irrelevant."

— DAVID OGILVY[1]

In his book *The Art of Thought,* Graham Wallas, a cofounder of the London School of Economics, points out that creative ideas are born in four stages: preparation, incubation, illumination, and verification.[2]

A demonstration of how this might play out can be found in Nobel Prize winner Kazuo Ishiguro's account of how he wrote his novel *The Remains of the Day* in just four weeks, revealed in a 2014 interview with the *Guardian* newspaper.[3] What's notable about this example is not just the astonishing speed at which he completed the novel but the way his creative process unfolded long before he put

pen to paper, and how his journey was a mosaic of distinct
mental states.

- **Preparation:** First, Ishiguro learned everything he
  could about the setting of his novel. As he describes
  it, he voraciously consumed "books by and about
  British servants, about politics and foreign policy
  between the wars." This would have been optimal in
  a high-energy gear 2 state.
- **Incubation:** He then put his project aside and filled
  his time with various social engagements for almost
  a year. Ishiguro was not actively doing anything to
  prompt his ideas to incubate during this time; they
  marinated in his subconscious mind.
- **Illumination and verification:** Ishiguro eventually
  locked himself in his home and drowned in the
  imaginary world of his future novel. He wrote furi-
  ously and with such energy that he penned many
  "scenes that went nowhere" while also having "vital
  imaginative breakthroughs." This was likely a high-
  energy gear 2 state which enabled a form of brainstorming, or in his words, "The priority was simply
  to get the ideas surfacing and growing." During this
  stage, where his "fictional world was more real [to
  me] than the actual one," he would have withdrawn
  his attention from the outside world and guided it

inward, into the imaginary "Darlington Hall" of his novel, where his mind now resided. After four weeks spent in this existence, *The Remains of the Day* took form.[4]

THE CREATIVE PROCESS

Whether you are trying to solve a complicated problem, think up a new idea, or write a novel, your mind is driven by a *process* and by a *goal*. Your goal is to hunt for that new idea or solution. In the course of pursuing that goal, your mind must sometimes widen its beam of attention to capture details that do not at first appear directly relevant.[5] If your attention is overly focused on your goal, it makes you less adventurous and narrows your thinking.

Earlier, I described how your attention widens in both the *lower* and *higher* energy states within gear 2. In each case, your focus becomes "leakier" and your mind can wander away from your immediate target, but with a nuanced difference: in the low-energy state, the intensity of your attention is weaker, whereas in the high-energy state it is stronger. This difference manifests as two subtly different kinds of creativity.

Spontaneous insights

At the less energetic end of gear 2, your mind is tranquil and relaxed. You can nudge it to focus, but you can also let it wander. Once you remove your attention from all anchors and let it gently wander and float, you cede the stage to your subconscious mind so it can incubate your thoughts and catalyse insights. This less energetic, slightly day-dreamy mental state primes your mind for *spontaneous creativity,* and you can follow spontaneous ideas through as soon as they appear by narrowing your attention at will. Letting your attention loose from its target also lets you gaze at the whole picture from afar and gain perspective.

Doing something that needs you to stay alert and demands some attention but does not strongly anchor your attention is one route to getting into this state of mind. This explains the abundance of anecdotes about people having moments of insight while out walking, jogging, showering, or doing the dishes but *rarely* when speaking at a meeting, watching the news, or working to a deadline.

Boredom and "idling" during dull meetings provide excellent opportunities to enter this state, as long as you are relaxed and your mind is not occupied by pressing worries or intrusive thoughts. The lethargy induced by the meeting can help your attention wander without leaving gear 2 because you have to maintain *some* focus on the meeting in case your voice is called for.

Thinking outside the box

At the more energetic end of gear 2, your mind generally feels sharper and faster. Here, too, you can nudge it into focus at will, but it is always ready to explore the unknown, driven by an inner energy. Your beam of attention is wider and your radar captures details you couldn't see when your focus was narrow. Solutions that were previously invisible now spring into view. The higher levels of norepinephrine in this state make your mind more able to let go of old ways of thinking and embrace new ones. You think more laterally, you bend the rules, and you can rapidly learn new rules as you go along. Once you have gathered your new ideas, you have to narrow your focus and look closely at each idea to see if it is any good, something you can do by virtue of being in gear 2.

This kind of *divergent* thought requires freedom from restrictions and rules. If you are too narrowly focused on your goal, or if you fear that unconventional or "out of the box" ideas will be dismissed or ridiculed, your divergent thinking skills diminish. Surprisingly, divergent thinking is even stifled by feeling physically constrained—for example, if you can't move freely or have to sit in a rigid position. In one experiment, raising the ceiling height by just 2 feet, from 8 feet to 10 feet, made people think more expansively.[6] This may explain why walking or running outdoors, where the "ceiling" is as high as the sky, can resolve mental blocks and inspire you with new ideas.

hyperefficient

Successful innovative thinking involves finding connections between seemingly disparate ideas. One example of this comes from Toshiyuki Murakami, the Japanese inventor of umbrella bagging machines. Ubiquitous at entrances to malls and hotels across Asia, these popular machines wrap your wet umbrella in a cellophane sheath to prevent it from dripping everywhere during monsoon rains. The genius of the machine lies in how quickly and smoothly the process happens.

In an interview with Japanese news channel NHK World, Murakami described how the idea behind the machine's mechanism was first inspired by a shoehorn. A shoehorn may not be something you would normally associate with an umbrella, but Murakami's creative mind connected the two items and used the shoehorn concept as the basis for the machine: inserting an umbrella into a bag is somewhat like inserting your foot into an obstinate shoe. During the development process, he came across a new problem: a cellophane bag, unlike an ordinary shoe, is notoriously difficult to pry open. He subsequently solved the problem by viewing an umbrella through the same lens as a letter. When you open an envelope, one edge is higher than the other; this lets you slide your letter in easily. Making the opening of each cellophane sheath asymmetrical so that one edge sits higher than the other lets you slide the tip of the umbrella in smoothly, just like a letter into an envelope.

Writing in 1981, psychology researcher Colin Martindale explained that "the more elements that a person can focus on simultaneously, the more likely that a creative idea will result."[7] One way to think about creativity is to imagine you have to rustle up dinner on a cold and wet Sunday evening. The wider the variety of ingredients you have in your kitchen cupboard, the more likely you are to come up with new ways to combine them into a spectacular recipe.

One striking example of how widening perspective helps with problem-solving comes from the Second World War. When the U.S. government commissioned a research group to find out where to place structural reinforcements on fighter planes, the research group initially approached the problem by looking at the existing data on where the returning aircraft were hit the most. They reasoned that the areas that sustained the largest number of bullet holes were most in need of extra protection. The mathematician Abraham Wald pointed out the terrible flaw in this logic. He suggested the researchers look at the data *they did not have:* the planes that hadn't returned. What if the planes with the bullet holes were able to fly back because of where their bullets struck them? This led to the realization that the parts of the plane that needed reinforcing were those that had no bullet holes in the planes that made it back.

Some research suggests that with practice, it may be

possible to improve your ability to see connections between disparate concepts by voluntarily widening your attentional net.

Researchers at Heidelberg University in Germany have shown that when players of a team sport are told to focus on a particular strategy, their attention narrows. If they play without a strategy, however, they spread their attention wider and approach complex situations with more creativity and "out of the box" flair. When a group of children practiced this "attention-broadening" strategy for one hour, twice a week for six months, they became better at using peripheral information to generate creative ideas.[8]

These principles can generally be applied to knowledge work: you are more likely to come upon innovative ideas and original insights when you widen your approach to narrow business objectives and immerse yourself in the creative process for its own sake.

The power of play

In a 1991 talk for Video Arts, British actor John Cleese described how, when Alfred Hitchcock hit a creative block while working on a screenplay, he would simply stop and "tell a story that had nothing to do with the work at hand."[9] At first, Cleese said, this infuriated Hitchcock's cowriters, but they soon realized that the point of the story was to dissipate the pressure everyone was under. The idea always surfaced when the pressure was gone.

Creativity is *impossible* in gear 3, so you are highly unlikely to come up with novel ideas if you are working under threat of judgment, blame culture, competition, or tension of any kind. Instead, ideas are most likely to germinate when your brain is guided by what it finds most pleasurable and is free to wander down new avenues without fear that they might lead nowhere. Play provides the perfect backdrop for this. Your mind shifts into a state of carefree experimentation and explores with abandon just like a child.

Play shares many characteristics with the core approaches to knowledge work that I have described throughout this book. First, play is intrinsically motivated — driven by curiosity, pleasure, and other inner rewards. Second, play can induce a flow state; think back to Tetris, or to Jonny Wilkinson's experience on the playing field. The philosophy behind the gamification of workplaces is predicated on these characteristics. Play also adds an element of fun to any workplace.

ECOSYSTEMS FOR INNOVATION

Claude Shannon wrote his landmark paper that spawned the information age while working at AT&T Bell Laboratories, in New Jersey. Arguably the greatest epicenter of creativity to exist in modern times, Bell Labs gave birth to the transistor, laser, charge-coupled device, and photovoltaic cell. Although the institution handpicked the most elite thinkers in each field, the secret to its extraordinary creativity may have been not merely the talented people who worked there but the architecture of its unique setup.[10]

Bell Labs was an industrial research laboratory devoted to inventing technology and crafting solutions to real world problems that could be fast-tracked into a manufacturing pipeline. The resulting breakthroughs often had a world-changing impact, giving researchers a palpable sense of meaning and agency. The director of Bell Labs during Shannon's time there, Mervin Joe Kelly, was guided by the notion that while leadership and teamwork are important in an organization, "it is in the mind of a single person that creative ideas and concepts are born." Kelly therefore sought to sculpt the optimal ecosystem for manufacturing such ideas. He viewed basic research as a precious reservoir—a "non-scheduled" zone in which researchers had the freedom to work at their own rhythm and pace, unshackled by deadlines or by the anxiety of financial concerns. Funding for research projects was guaranteed; a researcher's salary was dissociated from their research and instead linked to a merit-based ranking system, which inspired healthy competition. There were no short-term objectives demanding a continuous, linear output, so researchers could focus on long-term objectives without having to waste time demonstrating short-term progress to justify the project's funding.

The physical layout of Bell Labs was also conducive to idea generation, giving researchers the opportunity to alternate between convergent focus and divergent thought. If you wanted to focus, you could seal yourself into your own private office. But if you were stuck on a problem or needed inspiration, you could take a walk in the long corridor outside your office, where you might encounter a researcher from a different field or get a chance to peek into someone else's lab: serendipitous meetings that could nudge you into a fresh perspective or help hatch an *a-ha!* moment. A lunchroom behind a large, light-filled atrium provided a common space for all researchers to socialize in and encouraged the cross-fertilization of ideas and knowledge. The unfettered atmosphere within Bell Labs also gave researchers the freedom to express their diverse, creative, and often quirky interests: in one iconic photo of Shannon, he rides a unicycle across a corridor.

In these ways Bell Labs was much like a greenhouse for ideas, only instead of housing rare and extraordinary plants, nurturing them with the highest quality nutrients, and letting them cross-pollinate and grow, Bell Labs housed the best researchers in the field, nurtured them with a a sense of meaning, agency, and creative freedom, and let their ideas cross-pollinate and grow into novel and emergent solutions.

This kind of "greenhouse" model is better for idea generation than the traditional results-based, target-oriented, deadline-driven work practices in wide use today that are at odds with the rhythmic output of the creative thought process. But the greenhouse model can be financially impractical to implement fully. One way around this is to have a rotating, or "rolling," creativity team, in which every employee gets to spend a set length of time. During their time in the team, members work on a problem in a Bell Labs–style ecosystem, with a meaningful incentive to solve the problem but without demands, deadlines, or expectations for short-term progress. Managers explain the problem, provide resources, and simply let their workers be.

SOLVING DYNAMIC PROBLEMS

The challenge of solving problems in the AI-driven knowledge age is that problems are *dynamic,* meaning that the goals and obstacles are constantly changing. Familiar problems can suddenly transmute into entirely novel conundrums, so that whatever solution you are working on can become obsolete almost overnight.

If you can immerse yourself into the pleasure of the *process* of solving the problem, and not think too much about the *result* of solving it, you're more likely to sustain the mental stamina required for multistage complex problem-solving. Otherwise, your motivation can quickly turn into frustration as the shifting goals make you feel rudderless. When you can derive pleasure from the process of solving a problem, shifting goals and roadblocks make things more interesting and enjoyable. Intrinsic motivation can help you enter a "high energy" gear 2 state, and this state lets you access the set of skills that are most helpful for this kind of problem-solving. It allows your brain to focus while being flexible at the same time, so you can quickly identify and adapt to shifting goals. A high energy gear 2 state is also perfect for fluid reasoning.

Fluid and crystallized thinking

A problem can be familiar or unfamiliar. When you face a problem that resembles one you have previously encountered, you tend to draw on *crystallized intelligence*—that is, you use knowledge, information, and skills you already possess. An example of this is designing software that is similar to software you have designed before.

When the problem does not look like anything you have ever come across, you use *fluid intelligence,* which relies on logic, experimentation, and reason. For example, if you've never cooked before and need to bake a cake on a desert island without internet access, you will use fluid intelligence.

A focused gear 2 mindset helps with crystallized intelligence because it allows you to focus your attention on the information and ideas you already have. If you shift upward slightly into a high-energy gear 2 state, this primes you for fluid reasoning. It configures your brain to be malleable and plastic so you learn at an accelerated rate, and it widens your attentional scope so you think out-of-the-box. When you're operating in novel, hitherto unknown territory using fluid reasoning, you need the motivation to pursue the unknown, the attention to spot details hiding in the shadows, and the flexiblity to rapidly discard old rules and learn new ones.

As machines continue to encroach on human brainwork, fluid intelligence will become increasingly valuable—particularly in a world where the pace of technological change generates dynamic problems unlike those you have ever encountered before.

A TEMPLATE FOR CREATIVE PROBLEM-SOLVING

As you go about solving a problem with creativity, you hop across a patchwork of mental states, each state giving you the right tools for the specific phase of your journey. The journey itself requires changes in speed, the convergence and divergence of focus, and a retreat from and entry into the outer or inner world.

The Creative Mind

The first stage of solving any problem is to clear the mind of everything. The ideal mental state for this is gear 1, where your attention is too weak to dwell on anything for long and floats without commitment. You are more likely to be in this "offline" state first thing in the morning and last thing in the evening—which, anecdotally, are the two most popular times of the day for creative thinking.

For the next stage, it is helpful to zone into exactly what it is you are trying to accomplish and reduce the territory of the problem down to its core. Every superfluous detail creates a diversion and makes your signal noisy. To do this, you must narrow your focus and concentrate it on the center of the problem. This is best done in gear 2, so if you are starting from gear 1, adding a gear-raising activity like listening to fast-paced music or doing a bout of exercise before this phase and following it with the Quiet Eye technique—or a version of it, such as a short, focused-attention meditation session *immediately before* you begin—can help you reach the right state of mind.

As your journey subsequently unfolds, you have to decide when to think passively and when to think actively. If you hit a wall and cannot see a way through, it might help to ease your foot off the pedal and let your mind wander into a low-energy gear 2 state. Here, your attention partly floats and partly focuses; your perspective widens and your subconscious mind can lend a hand. Since you are still in gear 2, you can quickly zone into new ideas or insights and pursue them. Taking a walk or doing something that doesn't obligate you to think can inspire this state. At other times, you will make more progress by active thinking, for instance when you want to brainstorm ideas or actively pursue a chain of thought that leads you into uncharted territory—you will make more progress by active thinking. A quick bout of physical movement can help set the scene for this high-energy gear 2 state. Occasionally, you will run into barriers that seem impossible to overcome. At such times it may be best to leave your work completely, let your subconscious chip away at the problem in gear 1, and return to it when you are mentally refreshed.

If another kind of creative activity comes more easily to you than the creative task you are working on, doing it will prime your mind for your specific creative task.[11] According to his wife, Elsa, Albert Einstein often used this strategy. When Einstein was deep in the throes of a problem, he would emerge from his study, sit at his piano and play a few chords, then return to his problem. The playing seemed to trigger insights, which he would quickly scribble down and then return to his study to consolidate. This unusual practice may have helped Einstein enter exceptional mental states of almost superhuman insight.[12]

THE UNIQUE CHALLENGES OF THE DIGITAL AGE

We have now laid the groundwork for efficient mental work. Starting with the principles of rhythmic working, we have explored ways to navigate through various mental states to optimize different genres of knowledge work. It is time to add one final layer to our game plan and adapt these strategies to the specific challenges of the AI-assisted information age.

Your mind is the new assembly-line worker: it processes information to manufacture knowledge products. As the *load* of information grows ever heavier and its pace *accelerates* with each successive iteration of technological progress, your mind must work exponentially harder. And it must do so through a new landscape of *uncertainty*, where familiar uncertainties are gone and are replaced by novel uncertainties with network effects.

hyperefficient

In this era of acceleration, we need strategies that lighten the load and liberate mental disk space. Improving the quality of information, erasing unnecessary urgency, and slowing the scale of work to more human time all act as powerful antidotes.

11

CREAKING UNDER THE INFORMATION LOAD

"All meaning alters with acceleration, because all patterns of personal and political interdependence change with any acceleration of information."

—MARSHALL MCLUHAN[1]

Over the past century, the speed of information transfer has risen to unprecedented heights, while the cost of transmission has sunk to virtually zero. It is faster and at times easier to ask an app if it is raining outside than it is to step outside and see for yourself.

In the setting of knowledge work, the amount of information your brain is processing at any time fundamentally decides your gear. If it has to work through a large volume of data, it raises its gear to cope.

hyperefficient

In 1880, more than thirty-three million messages were sent across telegraph lines in the United States over the entire year. In 2023, at least ten billion emails were sent over the internet in the United States on a single day.[2] Unlike a telegram that travels from one person to another, email has an enormous blast radius: one message can reach thousands if not millions of people at any given time.

Today all the data held by the vast expanse of humanity has been aggregated into a "superlibrary" that is embedded in a distributed network of interconnected humans: the internet. With a single keystroke, some bit of digital data — no matter how inconsequential — can be captured and pushed onto the shelves of the superlibrary. The library's collection is limitlessly eclectic, ranging from more "traditional" knowledge (the entire *Encyclopaedia Britannica* can be on your screen in an instant) to obscure and irrelevant information (someone walking along Hollywood Road, Hong Kong, learns that a volleyball has gone astray in the Cloche d'Or park in Luxembourg City). Any human with an internet connection can access the entire content of this gargantuan library in a split second, at minimal expense, at any time.

This colossal rise in information transfer presents three related problems: high information volume, low information quality, and insufficient attention to cope.

When information transfer is cheap, information quantity increases at the expense of information quality. And as the quantity of information grows, your ability to access the right information recedes. A single Google query yields an endless number of links where the answers are long and garrulous, with varying degrees of relevance and quality. The most advanced AI-powered chatbots, like ChatGPT, having been trained on the prodigious information floating on the internet, often serve up what is most ubiquitous, rather than what is most relevant.

Picking out the relevant from the irrelevant requires you to process the surplus irrelevance, and your mental resources creak under the strain.

LIGHTENING THE LOAD

In the 1980s, a group of researchers investigated the ways in which your *working memory,* a temporary scratch pad of immediately retrievable information, interacts with *long-term memory,* the knowledge that has already taken up residence in the library of your mind. This led to *cognitive load theory,* which says that whenever you come across new information, you put it on temporary hold in your working memory before gradually transferring it to long-term storage. Your working memory has limited capacity. When it is crushed by the high volume of incoming information, this doesn't just make it harder to remember things: it also makes it harder to pay attention, learn, solve problems, and formulate new ideas. Your entire cognitive machinery grinds to a halt if your working memory gets overwhelmed.[3]

The researchers behind cognitive load theory have explored several ways to lighten the load on working memory by adjusting how you process, organize, and consume information.

Remove redundancy

Information can be organized in a way that minimizes the work your mind must perform to process it. For example, if you have a graph of your company's output on one half of your screen and supplementary information on the other, your mind has to perform the extra step of fusing the two together. If you integrate the information before you try to process it, your mind has less to do. Organizing information with a notepad and pen, instead of with your mind (that is, writing things down instead of trying to remember them), also reduces the load. This strategy also applies to the *amount* of information you present to your mind at any time. If you're viewing a presentation and each slide contains both a picture and a written description of that picture, your brain is exerting effort to process the same information twice— and more effort still goes into realizing that the information conveyed by each is identical.

Thin out traffic

When you have a lot of information to consume, distributing it across different mediums can lighten the load on your mind. For instance, if you're studying for an exam, you could spend half your time listening to recordings of class lectures and half the time poring through your written notes, or if you're reading a long book, you could switch back and forth between the hardcover and audio editions.

CD, a first officer at a major airline, told me he prefers flying a Boeing 777 to other planes that rely more heavily on automated flight navigation systems. Although automated flight navigation has reduced mental load on pilots in various ways, he finds that having to rely on flight and navigation displays for information most of the time can place a strain on visual attention. CD enjoys flying 777s because they still retain features (the control column, the rubber pedals, the thrust levers) that let him physically "feel" what the aircraft is doing (the pitch and roll, the yaw, the thrust) without depending entirely on visual displays. This distributed form of processing makes the experience less mentally taxing.

Add texture and variety

Remember that your working memory is where you temporarily shelve new information until a place can be found for it in your vast mental library. If you can connect that novel information to something you already know or have experience with, it becomes easier to file it away. The more varied or texturally rich your information is, the more likely it is to contain elements that resonate with something you are already familiar with, which means that you can place it in your mind's library more quickly, freeing up that space in your working memory. This partly explains why it's easier to learn something if it is presented as a story rather than a list of facts. The storyline already links the elements together, so they are easier to file.

Loosen the goal

One way to lighten mental load is to make the goal you are working toward *feel* less rigid. For example, when solving a complex problem, your brain is constantly comparing where you are with where you ultimately want to be. This takes up precious working memory real estate. A simpler, less rigid goal frees up disk space.

Consider these two problems. Which one feels more mentally taxing?

1. If a car's speed is 2 kilometres per hour and it was completely stationary a minute ago, how far has it traveled?

2. A car's speed is currently 2 kilometres per hour. It was stationary a minute ago. Find out as much information as you can from these two facts.

In the first scenario, your working memory has to simultaneously remember the goal and perform mathematical calculations. In the second example, your working memory does not have to remember the goal, so it has more space available for the calculations. Both approaches lead to the answer, but the second question will *feel* easier.

This principle also applies when you are learning a large body of information. In one experiment, a group of people pretended they were art thieves as they explored a virtual museum in a computer game. Half had to imagine they were *carrying out* a heist; the other half had to imagine they were *planning* a heist. Those who were planning a heist could explore the museum without having to keep the goal (to get the most expensive paintings) at the top of their minds, and were able to remember more of the paintings the next day.[4]

SIFTING THROUGH NOISE TO GET TO THE SIGNAL

As information quantity increases at the expense of information quality, the act of passing on information has become the goal, whether the person receiving the information needs it, understands it, or can verify its accuracy. It is easier to respond to an email with all the information you have on a given topic than it is to parse the sender's question and figure out what information they are specifically seeking. This encumbers them with *more* instead of less mental work as they now need to trawl through leagues of unnecessary information to arrive at the bit of information they actually need.

In 1986, researchers Richard Daft and Robert Lengel put forward a theory (known as *media richness theory*) that some types of communication channels are better than others for ensuring that a complex message is understood. The greater the uncertainty and risk of misunderstandings, the "richer" the channel should be.[5] Face-to-face communication is high in "richness" because the person sending the information can correct gaps in understanding and eliminate ambiguity immediately with hand gestures, facial expressions, and other nonverbal cues. If the information being communicated is incompletely understood or misunderstood, both the receiver and the sender have to put in extra effort to fill in the blanks. By improving the quality of the information, we remove the extra load on our minds.

Take the example of PB, who manages a successful team in a large multinational company. He has noticed over the years that complex queries requiring precise responses are increasingly being sent via email instead of being discussed at in-person meetings. These emails are often vaguely and imprecisely worded, and recipients struggle to understand what is being asked of them. To cover all bases, they reply with superfluous and ambivalent information, much of which is irrelevant. The original sender of the email is befuddled by the reply but too self-conscious to admit it, so the query goes unanswered, leading to a bigger problem down the line. Meanwhile, the confusion, uncertainty, and extra work increase the mental load of all parties.

Emails are often left unread because the already overwhelmed reader does not have the cognitive stamina to sift through a haystack of information to find the needle. Important content in an email can often get so buried in logorrhoea that it escapes the reader's notice. The undue frustration this causes wastes resources, and the process of following up and chasing replies further adds to the heavy load.

Making decisions

One consequence of low-quality information is that it becomes harder to make good decisions.

A decision is a three-step process resembling what happens in a court of law. You gather evidence, your brain deliberates on the evidence, and a judgment is passed.

The quality of your decision ultimately depends on the quality of your evidence, and your evidence will be more reliable if you are neither biased nor rushed.[6] Once you have gathered enough evidence, you have to weigh it, explore it, connect dots, and think critically. This deliberation phase requires you to oscillate between a wide attentional field and a narrow one, so you cast your net wide and then zone into the signal. A high-energy gear 2 state ticks all these boxes.

The less time you have to form a decision, the more truncated this process becomes, and the evidence-gathering stage is most vulnerable to being trimmed away.[7] Without adequate evidence, your brain is more likely to use the knowledge it already has, as well as any biases it holds, to guide its judgment. Being in a gear 3 mental state also abridges this process, leading to biased decisions and snap judgments—sometimes to catastrophic effect.

Strong stimulants such as amphetamines can easily drive the user into a gear 3 mental state. In early 2003, a military hearing took place in relation to an incident in Afghanistan in which two U.S. Air Force pilots mistakenly bombed Canadian soldiers. According to an NPR report, the pilots (via their lawyers) claimed their judgment had been distorted by the stimulant drugs they had been taking: according to news reports, military doctors in the United States at the time prescribed small doses of amphetamines to pilots to help them cope with sleep deprivation (though the Air Force did not directly attribute the tragedy to amphetamine use).[8] The use of stimulants to enhance performance on the war front wasn't new: seventy-two million amphetamine tablets were purchased for British troops in the Second World War, and a derivative of amphetamine was also used in Nazi Germany.[9,10] Over two thousand years ago, Emperor Augustus, the first emperor of Rome, reportedly took a very different approach to warfare. It is said that coins minted during his reign carried the emblem of a crab holding back the fluttering wings of a butterfly and the words *festina lente* (Latin) or σπεῦδε βραδέως (original Greek), meaning "make haste, slowly" because Augustus believed that rashness was a costly flaw in a military commander and that one can achieve more by slowing down.

Since working under pressure can easily push you into a gear 3 state, industries that rely on high-stakes decision-making invest in training staff to rehearse decision sequences that can be performed in a crisis, without the need to think. Nuclear power plants, oil rigs, submarines, and large banks are a few examples of workplaces with emergency protocol drills for times of crisis.

MAKING DECISIONS

To make **wiser** decisions ➡

> BREATHING EXERCISE
> - Do a 5-2-7 pattern breathing exercise for two minutes right before making a big decision. (Inhale for 5 seconds, hold for 2 seconds, and exhale for 7 seconds.) This increased the percentage of correct decisions by 47 percent in one study[11]

To make **faster** decisions ➡

> PHYSICAL EXERCISE
> - Any mild-moderate (but not intense) exercise leaves you more alert and accelerates reaction time

To make **easier** decisions ➡

> - REDUCE your options
> - SPLIT the exploring part from the choosing part. (Thinking about the different options is less taxing than deciding on one option)
> - TAKE a break

Reducing decision fatigue

Decision-making is so exhausting that the fatigue it gives rise to has its own special name: *decision fatigue*. Decision fatigue builds with every new decision made; as it builds, your mental resources get depleted and your decision-making ability becomes impaired. It is as if your brain copes with haemorrhaging resources by shutting off the tap.

Decision fatigue makes you more risk-averse. This is

why bank workers refuse more loans if they haven't had a break in a couple of hours, and why nurses answering NHS helpline calls become increasingly cautious with every call they take.[12] In one analysis of four hundred calls taken by 140 nurses, the nurses were 5.5 percent more likely to recommend that a patient seek medical care after every patient they spoke with, amounting to about a 20.5 percent increase every hour, independent of how many hours they had been working.[13] The gradual decline in decision quality was directly related to the number of decisions they had made, not the total time spent working.

Here are three ways to reduce decision fatigue:

- Reduce the number of options you must choose from. The more options you have, the more mental energy is needed to process and evaluate each one in detail.
- Do the groundwork first and delay the actual decision. *Having to choose* is more tiring than *exploring* choices. That's why window-shopping is less mentally fatiguing than actual shopping, and selecting several options to be narrowed down later is less fatiguing than choosing a single option now.
- Take as many breaks as possible, each for as long as possible, to give your brain time to replenish its resources.

OVERCOMING THE LIMITATIONS OF ATTENTION

The fantastic advances in technology and communication have made information so cheap that *attention* has become expensive. The limiting factor of information transfer is no longer the generation or delivery of information, but the attention of the person receiving it. In the Palaeolithic era, we hunted for information (to point us in the direction of food, water, and shelter). In the information age, we become the hunted—the information hunts us.

Since our attentional resources are finite, the relentless siren song of novel information forces us to shift from serial processing (when you do one thing at a time) to parallel processing (when you try to do many things at the same time) in an attempt to keep up. The effort of disengaging from what you are doing, engaging with the new data, disengaging from that, and then reengaging with what you were doing, repeatedly, is mentally costly and creates extra mental strain.

Multitasking

Seasoned live television anchors are especially adept at processing two asynchronous streams of information at the same time. This happens when they listen to their producer through an earpiece while speaking to the camera. They don't attend to one stream first and then the other, but somehow seem to fuse both streams into one source of reality. For most of us in the knowledge work world, multitasking involves rapidly toggling back and forth between different tasks within a short period of time. Since different jobs need different degrees of focus and speed, switching tasks necessarily involves shifting focus and sometimes changing gears.

Changing tasks is a little like changing relationships: you need to get over the previous relationship before you can fully commit to a new one. Whether you are shifting from reading an email to preparing for a presentation, or from attending a meeting to brainstorming ideas, you have to disengage from what you are doing in order to engage with what you will do next. By disengaging well, you make the transition efficient and less taxing. You disengage simply by unplugging your attention from your target, using the techniques described earlier.

The need to disengage becomes more important the more focused you are on what you are switching away from. Using the relationship analogy, you need more effort to move on from a partner you were deeply in love with. This is why multitasking is antithetical to deep, focused work: the more focused your attention is, the harder it is to disengage.

It is easier to disengage your attention if you are in a high-energy gear 2 state, when your focus is slightly porous and, as I described earlier, when your brain is configured to be more flexible and exploratory. This helps explain why a bout of energetic exercise right before you begin a multitasking session can help you perform better. In 2019, a Japanese study found that half an hour of moderate-intensity (treadmill) exercise improved a multitasking session performed about half an hour later.[14]

Sometimes a switch in tasks can require a rapid reset of your gear. This is particularly common in sport. If you are playing basketball or football (soccer), gear 3 fuels you to race across the pitch, and gear 2 gives you control when you score. Biathletes have to quickly flip from skiing downhill at maximal power in gear 3 to aiming the rifle with a steady eye in gear 2.

Multitasking is also becoming increasingly common in the knowledge work landscape. For example, someone responsible for the digital infrastructure of a company might need to oscillate between dropping everything to urgently fix a malfunction one moment and working on complex software with deep concentration the next, multiple times throughout the day.

It is considerably harder to downshift from gear 3 to gear 2 than it is to shift upward from gear 2 to gear 3. The urgency of the situation — and your emotional reaction to it — jerks your gear upward and helps the transition from

2 to 3, but you need to exert some conscious effort in order to shift your gear in the opposite direction. I have described strategies for doing this earlier in the book.

LIVING IN A VIRTUAL WORLD

As technology has raced forward, it has transmuted the solid, tactile world we inhabit into a virtual one that moves a thousand times faster. Our brains must navigate this virtual world with imagination, and they must do so at the speed of light.

In 2005, philosopher Ned Kock, a professor of information systems at Texas A&M International University, theorized that because humans have evolved to communicate with others while looking at them, listening to them, intuiting their nonverbal signals, and responding immediately, evolution has endowed us with specialized brain circuits that enable us to do these things without much effort. Once we start to communicate in less "natural" circumstances, where we can't see someone's face, hear their voice, read their nonverbal signals, or respond in real time, we are forced to use newer, less efficient brain circuitry—which is why digital communication feels more mentally taxing. Kock's research has shown that people expend between five and fifteen times more effort to convey complex ideas by email than by face-to-face communication.[15] The further we drift from the way we have evolved to communicate, the greater the load on the mind.

This is not the only way that our brains are better adapted to the tangible world over the virtual one. Research shows that reading from a tablet or screen, for example, imposes a heavier mental load on the mind than reading a physical book.[16] When you read a book, you have an implicit sense of information existing tangibly in space. You can feel your way back to a forgotten detail, a lost name. Time, space, and knowledge are tightly braided together. As the unread part of your book shrinks, you know the end of the story is near. There is no uncertainty about where you are and where you are headed. A tablet or e-reader robs you of the ability to carve the book's story into your physical space in this way. When you need to retrieve information, you have no physical cues for where the information is located. Your working memory has to do all the work.

Virtual reality is one of the more drastic examples of how modern-day work is shifting from the hands to the head. In the real world, your mind shares the workload with your body. Whether you are playing an instrument, building a prototype, or putting data into a spreadsheet, haptic cues give texture and context to your mind's model of reality and reinforce your brain's predictive processing. With a VR headset, this sensory information must be filled in by the mind.[17] This increases mental load and can hamper performance as well as some types of learning.[18]

12

TOO LITTLE TIME

"A nervous man cannot take out his watch and look at it when the time for an appointment or train is near without affecting his pulse."

——GEORGE MILLER BEARD, MD, 1881

In 1881, nerve specialist George Miller Beard lamented that the telegraph was partly to blame for the rising tide of nervousness in America. "Before the days of Morse and his rivals," Beard wrote, "prices fluctuated far less rapidly, and the fluctuations which now are transmitted instantaneously over the world were only known then by the slow communication of sailing vessels or steamships...whereas now, prices at each port are known at once all over the globe. This continual fluctuation of values, and the constant knowledge of those fluctuations in every part of the world, are the scourges of businessmen."[1]

By modern standards, the telegraph is archaically slow. Today fluctuations in prices and trades are transmitted at the speed of light. A hundred and thirty years after George Miller Beard lamented the telegraph, U.S.-based telecommunication company Hibernia Atlantic announced that it would slice five milliseconds off the time it takes to relay financial data between London and New York, thanks to a new high-speed undersea fiberoptic cable.[2] When the world trades at high frequency with automated algorithms, a thousandth of a second head start can mean a profit of millions.

When you shine light into your eyes, your pupils constrict at a speed considered "instant" by human standards. But that speed is actually a languorous fifth of a second, millions of times longer than iX-eCute, a purpose-built microchip that prepares trades in 740 billionths of a second, takes to execute an entire trade. It has been said that the global financial markets spin the world on its axis. Today the world is spinning so fast the human eye can no longer see it move.[3]

Your brain is always trying to match its own pace with what it *perceives* as the pace of the world so it can keep up with the stream of information flowing in. When you have to race to keep up with inflowing information, the world seems to be speeding ahead. Your perception of time plays into this dialogue. If you think you have less time in which to do something, your brain cranks up its gear; if you think you have ample time, it downshifts.

THE PROBLEM WITH BREVITY

The way we communicate with one another in the absence of technology is slower than how we communicate in its presence. Our pace of speaking, for example, is slower than our pace of speed-reading text. Long-form conversation is more leisurely than rapid texting or emailing. Pressed to relay information immediately, we no longer present information in context-rich narrative forms. The prose we once composed in letters decayed into sentences when email entered the scene. These sentences became phrases with instant messaging applications, and have now evolved into mere words, abbreviations, and emojis in text messages.

The pace of our information age favors stripped-down, lean facts over analysis and storytelling. We might assume that the shorter the message, the less information we have to process and remember, but in fact, this extreme brevity can make it *more* difficult to learn and grasp information. A recent research paper shows that we find fragments of data more difficult to learn when they are communicated as statistics or stand-alone facts, compared with when they are embedded into a story. Productivity gurus will tell you that "less is more," but in fact, less if often more mentally taxing.[4]

Making sense of the world takes time. If information is coming at you at breakneck speed from all directions, you are forced to press your foot on the pedal and ratchet up to gear 3 in order to keep up. But as we have learned, gear 3 is not conducive to anchoring your attention to a new piece of information long enough to be able to digest it; in gear 3, your brain is hypervigilant, ready to move on to the next thing even before the old thing has had a chance to register. In his book *The Information Society*, Robert Hassan writes of how the "networked computer with speed built into its capacity and design...promotes a kind of restlessness, a 'fluttering' whereupon the user is unable or unwilling to linger for too long." Thanks to "the volume of information we are confronted with and the time constraints social acceleration places upon us," we are forced to resort to "abbreviated thinking."[5] Ironically, these shortcuts make mental work inefficient; though a three-word text message may take less time to compose and read than a three-sentence email, discerning the text message's intended meaning drains many more resources.

THE PROBLEM WITH URGENCY

The more urgent information seems, the more compelled you feel to process it quickly. Urgency demands speed.

If you receive information at the very instant that information has come into existence, that information feels disproportionately urgent because you don't know its implications and your brain scrambles to resolve the uncertainty. This is why an item reported as "breaking news" feels more viscerally alarming than if it were reported in a routine news slot.

When you learn of an event the moment it takes place, you don't just feel close to it in time; you also feel close to it in space. The shorter the time between the event occurring and your receiving information about it, the stronger the illusion that you are directly experiencing it, even if it is happening far away. When you shrink time, you also shrink space.

The ease and speed of modern-day information transfer means that all information is almost instantly communicated, regardless of its importance or relevance. Most information is embellished with features that make it *look* urgent, even if it is not. This has created a sort of inflation in information in which everything is breaking news and you are being drowned in a tsunami of a hundred equally urgent emails every hour. To process all this information at the speed it demands (that is, instantly), the brain pushes down on the pedal to take itself to gear 3.

KR, a senior manager at a blue-chip company, recently told me how her attempts to manage the influx of emails have been stymied somewhat comically over the years. First, she tried prioritizing emails that were sent with a "high importance" label, but once senders realized that emails marked with that label would be answered faster, the vast majority of emails were labeled "high importance." Next, she made it known that writing "urgent" in the subject field would ensure that important emails were read. Over time, however, every other email carried an urgency indicator, making it impossible for KR to respond to all of them in a timely manner. In response, senders started repeating the word "urgent" in the subject field, leading to "urgent urgent" and "urgent urgent urgent," often followed by a flurry of exclamation marks and asterisks. As an additional tactic, senders began to loop more and more people in on a chain in order to elicit quick responses from "someone." Ironically, this manoeuvre resulted in emails going unanswered for even longer, because everyone assumed that it was someone else's duty to reply. KR told me that she has now come full circle: she finds that she pays more attention to an email if the action required is in the first few words of the subject line and the word "urgent" is not used at all! Aside from making emails harder to prioritize, the ubiquity of the word "urgent" places us in a permanent state of vigilance and anxiety, making it harder to shift down to gear 2 when we need to get focused work done.

THE PROBLEM WITH DEADLINES

The brain does not work at a consistent, fixed speed like an assembly line. Its pace ebbs and flows, and its attention waxes and wanes, so its output emerges in spurts.

Since the brain's work does not proceed at a constant rate, it cannot really be tracked by objective time. Your perception of time changes as you work. Think of how fast time passes when you are in gear 2 and "in the zone," and how it drags when you are in gear 1 and bored.

If you track mental work with objective time, the pace of time assumes false equivalence with the pace of work. Thirty minutes spent focusing intensely on an assignment is deemed equivalent to thirty minutes spent in a pointless meeting. Frequent deadlines encourage this false equivalence.

Deadlines also endanger creativity. If the anticipation of a deadline pushes you into gear 3, you cannot create. Remember: the extraordinary creativity at Bell Labs was made possible, at least in part, by not shackling researchers to deadlines.

THE PROBLEM WITH ATTENTION

By dissolving the boundaries of time and space, instant, around-the-clock communication has flattened work into a long march of continuous productivity for which we must maintain the same intensity of attention for long periods of time.

Work that demands continuous attention comes in two flavours. The work can be *active,* such as if your job requires you to bat away a never-ending barrage of problems that come at you one after the other, or *passive*—for example, if you are a supervisor watching for mistakes while an intelligent machine works.

Intense, continuous, *active* mental work is tiring, and performance can sink in as little as five minutes. Some researchers have suggested that this is why we experience a "mental block" when our brain's capacity is overwhelmed, and our brain conserves resources by briefly halting all activity and taking a minuscule rest. Proactively pausing for as little as five seconds every couple of minutes can help to sustain performance in this kind of work.

Increasingly, in jobs across different industries around the world, the role of *doer* is transforming into the role of *supervisor* as an intelligent machine does the work while a human continuously watches. If the system is efficient, errors are rare, so the supervisor watches passively for long hours, and boredom inevitably creeps in.

Coping when work demands passive vigilance

It can take as little as twenty minutes for your performance to slide in this kind of work.[6] After this time, your mind starts to wander, suggesting your brain is switching into gear 1, because you are either bored or tired and it is wise to take a break.[7] The optimal break duration for passive continuous work is three to ten minutes; if you take a longer break, your brain risks losing momentum. Generally, how often you take a break is more important than how long your break lasts.

Any passive work that needs you to stay continuously vigilant has an obvious disadvantage: it has no opportunity for Learning Progress. There is no tangible goal that you can work toward, no feedback on how you are progressing, and your effort to sustain attention is not consistently rewarded. If you are an air traffic controller watching out for possible accidents and things progress smoothly, you haven't actively done anything at the end of the day. You get no palpable feedback on anything if you are just watching the scene. In other words, the reason why such work feels so laborious despite demanding little mental energy is that none of the criteria for *intrinsic motivation* exist.

Despite being intolerably dull, this kind of work is often of critical importance, as in the case of a security guard monitoring security-video feeds for an intruder, or a scientist tracking real-time weather data for signs of impending catastrophic events. The more laborious the work feels, the more likely you are to be bored and distracted, which can cause you to miss red flags.

In 1943, the Royal Air Force asked a British psychologist, Norman Mackworth, to find out how long airborne radar operators could sustain their attention before they began missing target submarines. Mackworth created what came to be known as the "Mackworth Clock Test." If you were taking the test, you would have to watch a black pointer on a white background as it mimics the second hand of a clock, moving round and round a dial at a regular pace. Every now and again, the pointer jumps *twice* its usual distance; your ability to spot these jumps would be taken as measure of your vigilance. Most people sitting the test start missing the jumps after about thirty minutes.[8]

Organizations around the world are devising ways to use different kinds of brain-monitoring technology to help keep workers alert. Workers at the State Grid Zhejiang Electric Power Company in Hangzhou, China, for example, wear hats wired with sensors that measure their brain waves and convey warning signals when a worker is not in the right mental state for deep focus.[9] Supervisors then use that data to ensure that the distracted worker isn't being dispatched into high-risk situations, as well as to tailor breaks with precision so that tired workers get more rest than fresher ones. These methods increased profits by $315 million in just four years.

Since some of the drudgery of passive, continuous work stems from the absence of intrinsic motivation, actively introducing two of its ingredients—feedback and goals—can make the work more interesting.

Receiving feedback makes workers demonstrably more engaged.[10] The world feels more real and interesting when you speak to it and *it speaks back*. On this basis, embedding some kind of perceptible result of your action into these kinds of passive work—even something as insignificant as hearing the mouse click when you shift its position—can make a difference. In one experiment using an air traffic control simulation, the simple act of clicking on the screen to acknowledge flights entering the airspace (instead of just watching them) improved performance.[11]

Anchoring onto a goal can help you focus attention for long periods.[12] This explains the many anecdotes of people who navigate around deficits in attention by setting themselves goals and to-do lists, sometimes in the form of Post-it notes dotted around their computer screen or decorating their home. In 2011, California researchers tested a neuro-rehabilitation protocol known as goal-oriented attentional self-regulation (GOALS) training on a group of military veterans with chronic traumatic brain injury who struggled with controlling their attention. The protocol involves performing frequent "self-checkups" on whether you are still following your goal or are being distracted away. If you find yourself straying, you can actively bring your attention back with the mantra "Stop, relax, refocus": stop what you're doing, disengage your focus, and redeploy it on your goal. The military veterans showed significant improvements in their ability to pay attention after just five weeks of following the protocol.[13]

The monotony of passive, vigilant work slides you into gear 1, which makes it even harder to pay attention and carry on working. A final strategy to help with monotonous, passive, vigilant work is to use some kind of stimulation to surprise yourself into raising your gear. For example, when patients with damage to the blue dot network struggle with paying attention, simple things like clapping loudly or hearing a loud noise at regular intervals can help keep them on track.[14] Multitasking can also help, as long as the tasks are not complex. By demanding more resources, it forces your gear upward.

13

AN UNCERTAINTY PARADOX

"Il n'est pas certain que tout soit incertain."

—BLAISE PASCAL[1]

As technology has evolved, software has gradually replaced hardware in almost every facet of life in industrial countries, from toothbrushes and toasters to cars and operating theatres. In doing so, it has replaced things we can control with things we can merely *influence.*

Back when I had a 2002 VW, I could confidently look under the bonnet if it ever broke down and diagnose the problem. Today when my car decides to suddenly stop in the middle of a busy street, I am left flummoxed. I must take the car to the garage, where an engineer will "fix" the software. I can easily *influence* the car—summon it to move, stop, and turn—but I can't *control* it. The same is true for my toothbrush, television, and even my doorbell.

They are all operated by software, and I cannot take them apart to repair them should they malfunction. We cannot control what we do not understand, and most people (including myself) don't understand the algorithms that automatically bring my car to a halt if I am about to collide with another vehicle, or that send a notification to my phone letting me know that someone is at my door when I'm not at home.

As control is quietly and insidiously supplanted by influence, it makes the world around us feel increasingly uncertain. The uncertainty-technology relationship is somewhat like an ouroboros. The human discomfort with uncertainty has propelled us to seek and destroy it whenever possible, and that, in turn, has twisted the wheel of civilization and progress. But the technology born out of this progress has generated its own uncertainty. Many uncertainties that we have eliminated made us stronger by incrementally improving robustness through repetitive exposure, whereas many of the uncertainties we have introduced make us weaker owing to how our brains respond to their unknowable consequences.

GOING BEYOND HUMAN SCALE

If you walk along the trading floor of a financial firm these days, you might be hard-pressed to identify it as such. There is no one screaming across the floor with a cell phone glued to each ear; there is no frenzied typing or cacophony of phones ringing. If Gordon Gekko were to visit a modern-day trading floor, he might mistake it for a library.

The algorithms now doing the trading in financial institutions around the globe are not only quieter than humans: they also cannot make human errors, and they are much, much faster. In changing the paradigm of work, however, technological tools like algorithmic trading transmogrify the coordinates of work into dimensions that overwhelm human scale. Algorithms today enable trading at the fifth decimal point. At the fifth decimal point, a currency's exchange rate is impossibly variable — it sometimes just reflects randomness or noise, making the trade more uncertain than ever. When combined with the meteoric pace of trading, the net effect of this uncertainty would be too much for the human mind to handle.

Technological tools like algorithmic trading also change the nature of uncertainty: they may remove uncertainty at one level, only to introduce it — and amplify it — in another. In their 2020 research paper, researchers Martin Hilbert and David Darmon noted how algorithmic trading "decreased uncertainty on the microlevel and increased uncertainty on the macrolevel," and speculated that the reason lay in the interaction between complexity and uncertainty.[2]

TAMING UNCERTAINTY

Some historians say that Angkor, the medieval capital of the ancient Khmer empire, became the largest preindustrial city in the world because its people tamed uncertainty at an impressive scale.[3] In the ninth century CE, the Khmer built an elaborate water-management network covering an area bigger than Berlin. This guaranteed enough water — and therefore rice — for a million citizens, even during drier years, thus compensating for the uncertainty of monsoon rainfall. Even as people flocked to the city from regions near and far, it remained orderly and vibrant, a hub for art and scholarship. The king, Suryavarman II, commissioned the vast temple of Angkor Wat, dedicated to Vishnu, which continues to stand today.

But according to some researchers, the very same innovation that transformed Angkor into a dazzling city of monumental proportions was also the cause of its eventual downfall. While the water-management network made the city mightily efficient for the status quo, all redundancy was shaved off, which meant that when the pattern of monsoon rains shifted, the water-management system could no longer sustain the rice fields, which a million people depended on. Over time, citizens had come to take their water supply for granted — to the point that they forgot what it was like to experience a shortage. So when the shortage arose, they lacked the resources to deal with it. Under pressure from this and some other factors, the city collapsed. In the process of taming uncertainty, *the people of Angkor had forgotten it existed.*

Repeated exposure to tangible and actionable uncertainties can make a system more robust. It keeps vulnerabilities in view, out of which grow contingency plans for when the unforeseen strikes. Every shock to the system highlights its weaknesses, so the system can be built back better.

If we eliminate uncertainty altogether, we forget how to cope with it at all. If you swim in a small pool regularly, you won't be overwhelmed in an ocean. But if you hardly ever encounter water, the shallowest pond will make you nervous. By this logic, deliberately introducing a bit of chaos into a system can make the system more resilient to unexpected shocks when they arise.

INTRODUCING THE CHAOS MONKEY

Some years ago, Netflix engineers created a tool with such a philosophy in mind. In the Netflix technology blog, cloud engineers Yury Izrailevsky and Ariel Tseitlin used the analogy of a flat tire to describe their rationale for the project: "Imagine getting a flat tire.... Do you have the tools to change it? And, most importantly, do you remember how to do it right? One way to make sure you can deal with a flat tire on the freeway, in the rain, in the middle of the night is to poke a hole in your tire once a week in your driveway on a Sunday afternoon and go through the drill of replacing it."[4]

The team decided to apply that concept to their business operations and created a tool that simulated instances of systems failures. They called it "Chaos Monkey," for the way it mimicked a wild monkey running amok in their data center, chewing through cables and wreaking havoc. They explained, "By running Chaos Monkey in the middle of a business day, in a carefully monitored environment with engineers standing by to address any problems, we can still learn the lessons about the weaknesses of our system, and build automatic recovery mechanisms to deal with them. So next time an instance fails at 3 a.m. on a Sunday, we won't even notice."

In addition to making a system more robust and resilient to severe shocks, this approach has two benefits for your peace of mind. The first is you feel more secure knowing that you are *always prepared* for a worst-case scenario. The second is that, should a worst-case scenario present itself, you don't panic, because you already know how to approach it. Although the deliberate introduction of chaos temporarily shifts your gear up, it parks you at a lower baseline gear in the long run, while also preventing your gears from shooting up uncontrollably at the worst possible moment: when disaster has struck.

This type of chaos testing is increasingly becoming part of the standard IT protocol in many workplaces. For example, BB manages a team in a multinational corporation that is going through a digital transformation. His team has been integrating AI into a service that had previously needed frequent error correction. The AI made the service run like clockwork. There were no more errors, which meant that the team could leave the AI to run the service and devote their time to something else. This worked very well for some time, until one day the system crashed and needed urgent fixing. BB's team was caught completely off guard. It had been so long since they last had to fix the service's code that they had forgotten how to do it.

After this disastrous incident, BB introduced a new approach. The service would be intentionally and randomly broken on a regular basis, as a kind of "fire drill." This regular exposure to uncertainty would ensure that the team didn't lose the ability to deal with it, and would be ready for it whenever it presented.

When facing uncertainty, we can never know what the worst-case scenario is. We know the worst-case scenario that *has happened so far,* but this tells us nothing about what lies ahead. You cannot prepare for future uncertainty by assuming it will be of a particular form and magnitude and honing our skills accordingly. Instead, you are better off preparing for *any* uncertainty by training yourself to react in the most effective way. One way to practice this is by regularly exposing yourself to *some* volatility, ideally in controlled settings.

CONTROL AND SUPERSTITION

Our innate discomfort with uncertainty can be a significant obstacle to mental efficiency in a complex world. Worries and anxieties swamp our mental resources, divert our attention, and distort rational thinking, all by raising our gear. A sense of control and order can help to reduce the anxiety that stems from uncertainty, even if that control and order exist in a different domain. This explains why many organizational cultures that exist in an atmosphere of severe, unrelenting uncertainty have more rules of order and behavior than cultures that do not. Military culture across the world, for example, relies on strict hierarchical organization and high expectations of personal discipline. Maintaining perfect order in the way you make your bed and polish your boots the night before won't reduce the likelihood of running over a land mine tomorrow, but *psychologically* it gives you the illusion of having more control.

In the next sections, I will explore two approaches to mentally coping with uncertainty: training self-control and using rituals.

Training self-control

In my book *Stress Proof,* I described how your perception of a stressful experience depends on the intensity of your mental and physiological reaction to it. If you can make yourself stay calm during times of uncertainty, you experience less distress. Consequently, one of the most robust ways to prepare yourself for future uncertainties is by training your self-control.

Wingsuit BASE jumping has been called "the most lethal sport on earth." Wingsuit BASE jumpers "fly" through the air in a webbed, aerodynamic outfit at speeds well over 140 miles an hour—a velocity at which every micro-movement must be meticulously controlled, as the subtlest error is massively amplified and often fatal. Jeb Corliss, a master wingsuit BASE jumper who has relentlessly broken records over two decades in the sport, told me that it wasn't courage that led him to wingsuit jumping, but his desire to conquer *fear.*

At the tender age of six, defeating his fear of snakes became Corliss's first lesson in the power of self-control. Reasoning that if he could learn to control his emotional response in the face of his worst fears, he would be able to tame them, he began seeking out small snakes and forced himself to overcome his visceral terror in their presence. Once successful, he progressed to bigger and more dangerous snakes, always one step at a time. He followed a similar

strategy with BASE jumping: he "started small" and progressed through incrementally greater levels of danger. In
one of his most famous videos, Corliss manoeuvres himself
so precisely that he slices through fine balloon strings
barely metres from the ground. This extraordinary opera-
tion requires a gear 2 level of mental control, but the situation would place most ordinary people in gear 3. Corliss
says he was just as terrified as anyone else would have been
when he began his journey; the only difference was that his
self-regulation training had honed his ability to remain in
the right mental space despite mortal danger.

The U.S. Navy SEALs use this principle, too. A key element of SEAL training is to repeatedly expose trainees to
terrifying situations, so they get better at managing their
responses and staying in gear 2. One example is an underwater pool competency test that trains recruits to overcome
the panic induced by the threat of drowning. Recruits have
to stay underwater for twenty minutes while fighting off an
"enemy" (an instructor) who is trying to disable their
breathing equipment — and often succeeds. To overcome
the fear of being underwater without oxygen, recruits practice maintaining enough focus to be able to reassemble
their breathing equipment without their minds kicking
into a high gear.

In 2007, researchers meticulously interviewed eight
elite athletes, all of whom were either Olympic or world
champion gold medalists, as well as seven trainers of elite

athletes (three coaches and four psychologists) to understand how an athlete can attain *superelite* levels of performance. Of the key practices they identified, at least three targeted self-control:

1. Imagining every "terrible" scenario possible (without losing self-control)
2. Experiencing progressively increasing levels of mental pressure (without losing self-control)
3. Deliberate exposure to obstacles or failures (without losing self-control).

Most workplaces lack opportunities for training self-control. It is best to train it outside of work, through a hobby or pastime, with gentler pursuits such as hatha yoga training and tightrope walking or tougher activities like ultra-challenges and expeditions.

Using rituals

Superstitions and rituals can be powerful psychological "uncertainty reducers." As the British-Canadian psychologist Daniel E. Berlyne has written, "It is significant that vocations like those of the miner, the air pilot, the actor, and the professional sportsman, in which there are unremitting risks of disaster, whether from injury or death or from loss of reputation through poor performance, are notoriously conducive to superstition."[5]

If you take your sum total of uncertainty on a typical day, you can divide it into *controllable* and *uncontrollable*. A superstition is simply a belief that certain events are caused by forces outside your control. If you don't believe in superstitions, you carry the entire load on your shoulders. But if you *believe* in superstition, you can delegate the uncontrollable load to these unseen forces.

When you worry about an event in the past or in the future, your brain tries to decipher why it happened or what can be done to prevent it from happening. If you convince yourself that the event is completely out of your hands, your brain is less likely to waste time trying to influence it. The ultimate effect of delegating uncontrollable uncertainty is that it removes your attention from it. As a result, you are less distracted, you can sleep better at night, and you perform better during the day.

If you aren't the superstitious type, rituals can give you a similar sense of control over the uncontrollable. In 2012, Greg Garber, a senior writer at ESPN, described a twelve-step ritual that tennis great Rafael Nadal would follow before each serve.[6] These steps included sliding his right foot along the baseline to clean it, removing the dirt from his left shoe before his right one, adjusting the left shoulder of his shirt before the right, touching his nose and then his hair on the left side of his head, and then again his nose, followed by the right side of his head, and so on. Having some action or series of actions that you do every time you have to perform in a high-pressure situation (as when giving a speech or presentation) can help lower your gear and keep you focused. You don't have to believe that doing the ritual will magically improve your outcome; the sense of control comes from the consistency of doing those same actions each time, not from the actions themselves.

CONCLUSION:
FROM MARCHING SOLDIER TO
SPINNING DANCER

"Now in the electric age of... information, we begin to chafe under the uniformity of clock time. In this age... we seek multiplicity, rather than repeatability, of rhythms. This is the difference between marching soldiers and ballet."

—MARSHALL MCLUHAN, 1964[1]

In his 1964 work, *Understanding Media,* philosopher Marshall McLuhan declared that literacy, and subsequently the Gutenberg press, shaped the modern human tendency to process information in a segmented and serial way, which eventually contributed to the principle of the assembly line. But the acceleration of information transfer, beginning in the electric age, changed everything. Information could no longer be processed in a slow sequence of regimented steps. Life needed to be lived in rhythms.

As the landscape of knowledge work continues its tectonic shift sixty years later, accelerating time and inflating the flow of information to unprecedented dimensions, we can choose to respond in one of two ways. As the world accelerates, we can try to accelerate with it, staying in gear 3 for longer and longer periods in an attempt to keep pace. The limitation of this approach is machines will inevitably outperform us as they leap beyond human scale.

An alternative approach is to not compete at all; instead of trying to change the speed at which we work, we can *change the pattern* in which we work. We can shift, as McLuhan describes, from marching like a soldier in a straight line to spinning like a dancer moving in an undulating rhythm.

If we embrace a rhythmic way of working, we will be less affected by the pace of acceleration; a rhythmic axis is nonlinear. At the same time, we will create the best mental landscape for continuous learning, creative idea generation, and innovative problem-solving, while safeguarding opportunities for rest and rejuvenation. Choosing this second option removes the shackles imposed on our minds by unyielding assembly lines, and restores us to how we used to be, long before the rigidity of industrialization.

The tsunami of technological change is often viewed with trepidation because we fear it will cripple, exhaust, and outsmart human potential. Paradoxically, it may, in fact, do the opposite. If it pushes us into a rhythmic way of working, technological change will *restore* the human capacity to create and innovate that has long been suffocated by the imposition of the assembly-line template. The marching soldier's ossified mind from McLuhan's description will revert to one that dances with imagination and abandon. This is how we become hyperefficient—and win against the machines.

ACKNOWLEDGMENTS

Three remarkable women made this book possible. Thank you, Liz Gough, for believing in this book long before it took shape, for having faith in its author, and for giving the book its name. Thank you, Talia Krohn, for guiding the book's long journey with unfailing patience, immense wisdom, and meticulous attention to detail. Thank you, Andrea Somberg, for your abundant advice and support at every point in the book's trajectory.

Thank you, Dan Buettner, for opening my eyes to the astonishing world of the Blue Zones, and for the enormous privilege of visiting Seulo, Sardinia, for your Netflix project. Thank you to Mara Mather from the University of Southern California for taking the time to describe your findings on phasic LC firing, resonant breathing, and autonomic dynamics. Thank you to Pierre-Yves Oudeyer, research director at INRIA (Institut national de recherche en sciences et technologies du numérique), for taking the time to explain the Learning Progress mechanism to me. Thank you to Marcus K. Weldon, former President of Bell

Acknowledgments

Labs, for your time and patience in enlightening me about life at Bell Labs in the Shannon era.

I am deeply indebted to Jonny Wilkinson for revealing with great candor the mental gymnastics that takes place in the mind of an elite athlete. Thank you, Jeb Corliss, for taking the time to explain the mindset that undergirds the remarkable trajectory of your life. Thank you to the anonymous individuals who have generously shared their personal experiences and thoughts throughout the book. My thanks also to Pat Berach, Lisa Burke, Talia and Dan Ezra, Paul Hegarty, Ben Hodsdon, Alexandra Schuilwerve, and Benoit Ursino.

My deepest gratitude to my husband, Laurent, for your endless patience, enduring optimism, and everlasting love.

NOTES

A New Kind of Efficiency

1. McLuhan, M., & Powers, Bruce R. (1989). *The global village: Transformations in world life and media in the 21st century.* Oxford University Press.
2. McNamara, R. (2019, January 28). The volcanic eruption at Krakatoa. *ThoughtCo.* https://www.thoughtco.com/volcano-eruption-at-krakatoa-in-1883-1774022#:~:text=The%20eruption%20of%20the%20volcano,other%20islands%20in%20the%20vicinity
3. Railway time standards (1883, April 19). *The New York Times*, 1. https://www.nytimes.com/1883/04/19/archives/railway-time-standards-a-system-by-which-annoyances-will-be-avoided.html
4. O'Toole, J., et al. (1972, December). *Work in America: Report of a special task force to the Secretary of Health, Education, and Welfare.* https://files.eric.ed.gov/fulltext/ED070738.pdf

Chapter 1: *Power Laws*

1. Robert Lawrence Kuhn (host). Closer to truth — Roger Penrose Interviews. "Did the Universe Begin?" https://youtu.be/OFqjA5ekmoY?si=7ong1IdXV7IpQN9t
2. Ibid.
3. Suzman, J. (2017). *Affluence Without Abundance: The Disappearing World of the Bushmen.* Bloomsbury.
4. Aungle, P., & Langer, E. (2023). Physical healing as a function of perceived time. *Scientific Reports, 13,* Article 22432. https://doi.org/10.1038/s41598-023-50009-3
5. Sahlins, M. (2017). *Stone age economics.* Routledge.
6. Ibid.
7. Raichlen, D. A., Wood, B. M., Gordon, A. D., Mabulla, A. Z. P., Marlowe, F. W., & Pontzer, H. (2013). Evidence of Lévy walk foraging

patterns in human hunter-gatherers. *Proceedings of the National Academy of Sciences, 111*(2), 728–733. https://doi:10.1073/pnas.1318616111

8. Reynolds A., Ceccon, E., Baldauf, C., Medeiros, T. K., & Miramontes, O. (2018). Lévy foraging patterns of rural humans. *PLOS ONE, 13*(6), Article e0199099. https://doi.org/10.1371/journal.pone.0199099

9. Humphries, N. E., & Sims, D. W. (2014). Optimal foraging strategies: Lévy walks balance searching and patch exploitation under a very broad range of conditions. *Journal of Theoretical Biology, 358*, 179–193. https://doi:10.1016/j.jtbi.2014.05.032

10. Marshall McLuhan 1965 — The future of man in the electric age. Interview. https://youtu.be/0pcoC2l7ToI?si=46gw5f4fcfMlDG3K

11. Costa T., Boccignone, G., Cauda, F., & Ferraro, M. (2016). The foraging brain: Evidence of Lévy dynamics in brain networks. *PLOS ONE, 11*(9), Article e0161702. https://doi:10.1371/journal.pone.0161702

12. Patel, M., & Rangan, A. (2017). Role of the locus coeruleus in the emergence of power law wake bouts in a model of the brainstem sleep-wake system through early infancy. *Journal of Theoretical Biology, 426*, 82–95. https://doi:10.1016/j.jtbi.2017.05.027

13. Vázquez A., Oliveira, J. G., Dezsö, Z., Goh, K., Kondor, I., & Barabási, A. (2006). Modeling bursts and heavy tails in human dynamics. *Physical Review E, 73*, Article 036127. https://doi.org/10.1103/PhysRevE .73.036127

Chapter 2: *Your Gear Network*

1. Doyle, A. C., *The Complete Sherlock Holmes: Volume II.*

2. Of interest, the blue dot was first identified by Marie Antoinette's doctor, Félix Vicq d'Azyr.

3. Omoluabi, T., Torraville, S. E., Maziar, A., Ghosh, A., Power, K. D., Reinhardt, C., Harley, C. W., & Yuan, Q. (2021). Novelty-like activation of locus coeruleus protects against deleterious human pretangle tau effects while stress-inducing activation worsens its effects. *Alzheimer's & Dementia, 7*(1), Article e12231. https://doi:10.1002/trc2 .12231

4. Patel, M., & Rangan, A. (2017). Role of the locus coeruleus in the emergence of power law wake bouts in a model of the brainstem sleep-wake system through early infancy. *Journal of Theoretical Biology, 426*, 82–95. https://doi:10.1016/j.jtbi.2017.05.027

5. Gall, A. J., Joshi, B., Best, J., Florang, V. R., Doorn, J. A., & Blumberg, M. S. (2009). Developmental emergence of power-law wake behavior

depends upon the functional integrity of the locus coeruleus. *Sleep, 32*(7), 920–926. https://doi.org/10.1093/sleep/32.7.920

6. Jubera-García, E., Gevers, W., & Van Opstal, F. (2020). Influence of content and intensity of thought on behavioral and pupil changes during active mind wandering, off focus and on-task states. *Attention, Perception, & Psychophysics, 82*, 1125–1135. https://doi:10.31234/osf.io/3nr9v

7. Mather, M., Clewett, D., Sakaki, M., & Harley, C. W. (2016). Norepinephrine ignites local hotspots of neuronal excitation: How arousal amplifies selectivity in perception and memory. *Behavioral and Brain Sciences, 39*, Article e200. https://doi:10.1017/S0140525X15000667

8. Palagini, L., Moretto, U., Dell'Osso, L., & Carney, C. (2017). Sleep-related cognitive processes, arousal, and emotion dysregulation in insomnia disorder: The role of insomnia-specific rumination. *Sleep Medicine, 30*, 97–104. https://doi:10.1016/j.sleep.2016.11.004

Chapter 3: *What Is Your Gear Personality?*

1. Zheng, Y., Tan, F., Xu, J., Chang, Y., Zhang, Y., & Shen, H. (2015). Diminished P300 to physical risk in sensation seeking. *Biological Psychology, 107*, 44–51. https://doi:10.1016/j.biopsycho.2015.03.003

2. Strauß, M., Ulke, C., Paucke, M., Huang, J., Mauche, N., Sander, C., Stark, T., & Hegerl, U. (2018). Brain arousal regulation in adults with attention-deficit/hyperactivity disorder (ADHD). *Psychiatry Research, 261*, 102–108. https://doi:10.1016/j.psychres.2017.12.043

3. Williams, J., & Taylor, E. (2006). The evolution of hyperactivity, impulsivity and cognitive diversity. *Journal of the Royal Society, Interface, 3*(8), 399-413. https://doi.org/10.1098/rsif.2005.0102

4. Mundie, S. (interviewer). ADHD has been my superpower (2023, November 1) [Audio podcast episode]. In *Rethinking ADHD*. Qbtech. https://podcasts.apple.com/us/podcast/rethinking-adhd-qbtech/id1714485901

5. Catani, M., & Mazzarello, P. (2019). Grey matter Leonardo da Vinci: A genius driven to distraction, *Brain, 142*(6), 1842–1846. https://doi.org/10.1093/brain/awz131

6. Soares, P. S. M., de Oliveira, P. D., Wehrmeister, F. C., Menezes, A. M. B., & Gonçalves, H. (2022). Is screen time throughout adolescence related to ADHD? Findings from 1993 Pelotas (Brazil) birth cohort study. *Journal of Attention Disorders, 26*(3), 331–339. https://doi:10.1177/1087054721997555

7. Liu, H., Chen, X., Huang, M., Yu, X., Gan, Y., Wang, J., Chen, Q., Nie, Z., & Ge, H. (2023). Screen time and childhood attention deficit

hyperactivity disorder: A meta-analysis. *Reviews on Environmental Health.* E-pub ahead of print. https://doi:10.1515/reveh-2022-0262

8. Arvaneh, M., Robertson, I. H., & Ward, T. E. (2019). A P300-based brain-computer interface for improving attention. *Frontiers in Human Neuroscience, 12*, Article 524. https://doi:10.3389/fnhum.2018.00524

9. Barth, B., Mayer-Carius, K., Strehl, U., Wyckoff, S. N., Haeussinger, F. B., Fallgatter, A. J., & Ehlis, A. (2021). A randomized-controlled neurofeedback trial in adult attention-deficit/hyperactivity disorder. *Scientific Reports, 11*(1), Article 16873. https://doi:10.1038/s41598-021 -95928-1

10. Ziegler, D. A., Simon, A. J., Gallen, C. L., Skinner, S., Janowich, J. R., Volponi, J. J., Rolle, C. E., Mishra, J., Kornfield, J., Anguera, J. A., & Gazzaley, A. (2019). Closed-loop digital meditation improves sustained attention in young adults. *Nature Human Behaviour, 3*(7), 746–757. https://doi:10.1038/s41562-019-0611-9

11. This is based on the VIGALL test. Jawinski, P., Kittel, J., Sander, C., Huang, J., Spada, J., Ulke, C., Wirkner, K., Hensch, T., & Hegerl, U. (2017). Recorded and reported sleepiness: The association between brain arousal in resting state and subjective daytime sleepiness. *Sleep, 40*(7), Article zsx099. https://doi:10.1093/sleep/zsx099

12. Losert, A., Sander, C., Schredl, M., Heilmann-Etzbach, I., Deuschle, M., Hegerl, U., & Schilling, C. (2020). Enhanced vigilance stability during daytime in insomnia disorder. *Brain Sciences, 10*(11), Article 830. https://doi:10.3390/brainsci10110830

Chapter 4: *The Rhythms of the World*

1. Lawrence, D. H. (1993 [1929]). A Propos of *Lady Chatterley's Lover.* Cambridge University Press.

2. The measurements were taken from the fluid in the brain, where levels approximately reflect the extra norepinephrine floating around after it is released by the blue dot network. The measurements don't reflect the absolute total levels in the brain, which are heavily influenced by norepinephrine synthesis and other confounding factors.

3. Ziegler, M. C., Lake, C. R., Wood, J. H., & Ebert, M. H. (1976). Circadian rhythm in cerebrospinal fluid noradrenaline of man and monkey. *Nature, 264*(5587), 656–658. https://doi.org/10.1038/264656a0

4. Monk, T. H. (2005). The post-lunch dip in performance. *Clinics in Sports Medicine, 24*(2), Article e15-23, xi–xii. https://doi:10.1016/j .csm.2004.12.002

Notes

5. Hayashi, M., Morikawa, T., & Hori, T. (2002). Circasemidian 12 h cycle of slow wave sleep under constant darkness. *Clinical Neurophysiology, 113*(9), 1505–1516. https://doi:10.1016/s1388-2457(02)00168-2

6. Zhu, B., Dacso, C. C., & O'Malley, B. W. (2018). Unveiling "musica universalis" of the cell: A brief history of biological 12-hour rhythms. *Journal of the Endocrine Society, 2*(7), 727–752. https://doi:10.1210/js.2018-00113

7. Bes, F., Jobert, M., & Schulz, H. (2009). Modeling napping, post-lunch dip, and other variations in human sleep propensity. *Sleep, 32*(3), 392–398. https://doi:10.1093/sleep/32.3.392

8. Smith, A., Ralph, A., & McNeill, G. (1991). Influences of meal size on post-lunch changes in performance efficiency, mood, and cardiovascular function. *Appetite, 16*(2), 85–91. https://doi:10.1016/0195-6663(91)90034-p

9. Dutheil, F., Danini, B., Bagheri, R., Fantini, M. L., Pereira, B., Moustafa, F., Trousselard, M., & Navel, V. (2021). Effects of a short daytime nap on the cognitive performance: A systematic review and meta-analysis. *International Journal of Environmental Research and Public Health, 18*(19), Article 10212. https://doi:10.3390/ijerph181910212

10. Hayashi, M., & Hori, T. (1998). The effects of a 20-min nap before post-lunch dip. *Psychiatry and Clinical Neurosciences, 52*(2), 203–204. https://doi:10.1111/j.1440-1819.1998.tb01031.x

11. Conte, F., De Rosa, O., Albinni, B., Mango, D., Coppola, A., Malloggi, S., Giangrande, D., Giganti, F., Barbato, G., & Ficca, G. (2021). Learning monologues at bedtime improves sleep quality in actors and non-actors. *International Journal of Environmental Research and Public Health, 19*(1), 11. https://doi.org/10.3390/ijerph19010011

12. Valdez, P. Circadian rhythms in attention. (2019). *Yale Journal of Biology and Medicine, 92*(1), 81–92. PMID: 30923475; PMCID: PMC6430172

13. Baer, T., & Schnall, S. (2021). Quantifying the cost of decision fatigue: suboptimal risk decisions in finance. *Royal Society open science, 8*(5), 201059. http://doi.org/10.1098/rsos.201059

14. Wieth, M. B., & Zacks, R. T. (2011). Time of day effects on problem solving: When the non-optimal is optimal. *Thinking & Reasoning, 17*(4), 387–401. https://doi.org/10.1080/13546783.2011.625663

15. Taillard, J., Sagaspe, P., Philip, P., & Bioulac, S. (2021). Sleep timing, chronotype and social jetlag: Impact on cognitive abilities and psychiatric disorders. *Biochemical Pharmacology, 191*, Article 114438. https://doi:10.1016/j.bcp.2021.114438

16. Baehr, E. K., Revelle, W., & Eastman, C. I. (2000). Individual differences in the phase and amplitude of the human circadian temperature

Notes

rhythm: With an emphasis on morningness–eveningness. *Journal of Sleep Research, 9*, 117–127. https://doi.org/10.1046/j.1365-2869.2000.00196.x

17. Lara, T., Madrid, J. A., Correa, Á. (2014). The vigilance decrement in executive function is attenuated when individual chronotypes perform at their optimal time of day. *PLOS ONE, 9*(2), Article e88820. https://doi.org/10.1371/journal.pone.0088820

18. Yuda, E., Ogasawara, H., Yoshida, Y., & Hayano, J. (2017). Enhancement of autonomic and psychomotor arousal by exposures to blue wavelength light: Importance of both absolute and relative contents of melanopic component. *Journal of Physiological Anthropology, 36*(1), Article 13. https://doi:10.1186/s40101-017-0126-x

19. Tian, Y., Ma, W., Tian, C., Xu, P., & Yao, D. (2013). Brain oscillations and electroencephalography scalp networks during tempo perception. *Neuroscience Bulletin, 29*(6), 731–736. https://doi.org/10.1007/s12264-013-1352-9

20. Gomez, P., & Danuser, B. (2007). Relationships between musical structure and psychophysiological measures of emotion. *Emotion, 7*(2), 377–387. https://doi:10.1037/1528-3542.7.2.377

21. Shih, Y., Huang, R., & Chiang, H. (2012). Background music: Effects on attention performance. *Work, 42*(4), 573–578. https://doi:10.3233/WOR-2012-1410

22. Jafari, M. J., Khosrowabadi, R., Khodakarim, S., & Mohammadian, F. (2019). The effect of noise exposure on cognitive performance and brain activity patterns. *Open Access Macedonian Journal of Medical Sciences, 7*(17), 2924–2931. https://doi:10.3889/oamjms.2019.742

23. Belojevic, G., Jakovljevic, B., & Slepcevic, V. (2003). Noise and mental performance: Personality attributes and noise sensitivity. *Noise Health, 6*(21), 77–89. PMID: 14965455

Chapter 5: *The Rhythms of the Body*

1. Kerouac, J. (1990 [1958]). *The Dharma Bums*. Penguin Books.

2. Sir David Attenborough (writer and presenter). (2003, February 5). Persistence hunting, eight-minute segment of "Food for Thought" (episode 10) [TV series episode]. In Mike Salisbury (executive producer), *The Life of Mammals*. BBC. https://www.bbc.co.uk/programmes/p0009lwhq

3. Yamazaki, Y., Suwabe, K., Nagano-Saito, A., Saotome, K., Kuwamizu, R., Hiraga, T., Torma, F., Suzuki, K., Sankai, Y., Yassa, M. A., & Soya, H. (2023). A possible contribution of the locus coeruleus to arousal

enhancement with mild exercise: Evidence from pupillometry and neuromelanin imaging. *Cerebral Cortex Communications, 4*(2), Article tgad010. https://doi:10.1093/texcom/tgad010

4. Zhang, W., Wang, X., Li, X., Yan, H., Song, Y., Li, X., Zhang, W., & Ma, G. (2023). Effects of acute moderate-intensity aerobic exercise on cognitive function in E-athletes: A randomized controlled trial. *Medicine, 102*(40), Article e35108. https://doi:10.1097/MD.0000000000035108

5. Mehren, A., Özyurt, J., Lam, A. P., Brandes, M., Müller, H. H. O., Thiel, C. M., & Philipsen, A. (2019). Acute effects of aerobic exercise on executive function and attention in adult patients with ADHD. *Frontiers in Psychiatry, 10*, Article 132. https://doi:10.3389/fpsyt.2019.00132

6. Mehren, A., Özyurt, J., Thiel, C. M., Brandes, M., Lam, A. P., & Philipsen, A. (2019). Effects of acute aerobic exercise on response inhibition in adult patients with ADHD. *Scientific Reports, 9*, Article 19884. https://doi.org/10.1038/s41598-019-56332-y

7. Drew, R. C. (2017). Baroreflex and neurovascular responses to skeletal muscle mechanoreflex activation in humans: An exercise in integrative physiology. *American Journal of Physiology-Regulatory, Integrative and Comparative Physiology, 313*(6), Article R654-R659. https://doi:10.1152/ajpregu.00242.2017

8. Mather, M., Huang, R., Clewett, D., Nielsen, S. E., Velasco, R., Tu, K., Han, S., & Kennedy, B. L. (2020). Isometric exercise facilitates attention to salient events in women via the noradrenergic system. *NeuroImage, 210*, Article 116560. https://doi:10.1016/j.neuroimage.2020.116560

9. Goldman, B. (2017, March 30). Study shows how slow breathing induces tranquility [Press release, Stanford Medicine News Center]. https://med.stanford.edu/news/all-news/2017/03/study-discovers-how-slow-breathing-induces-tranquility.html

10. Lehrer, P. M. & Gevirtz, R. (2014). Heart rate variability biofeedback: How and why does it work? *Frontiers in Psychology, 5*, 756. https://doi.org/10.3389/fpsyg.2014.00756

11. Bachman, S. L., Cole, S., Yoo, H. J., Nashiro, K., Min, J., Mercer, N., Nasseri, P., Thayer, J. F., Lehrer, P., & Mather, M. (2023). Daily heart rate variability biofeedback training decreases locus coeruleus MRI contrast in younger adults in a randomized clinical trial. *International Journal of Psychophysiology, 193*, Article 112241. https://doi:10.1016/j.ijpsycho.2023.08.014

12. Shattock, M. J., Tipton, M. J. (2012). "Autonomic conflict": A different way to die during cold water immersion? *Journal of Physiology, 590*(14), 3219-30. https://doi:10.1113/jphysiol.2012.229864

Notes

13. Rajkowski, J., Kubiak, P., & Aston-Jones, G. (1994). Locus coeruleus activity in monkey: Phasic and tonic changes are associated with altered vigilance. *Brain Research Bulletin, 35*(5–6), 607–616. https://doi:10.1016/0361-9230(94)90175-9

14. Lebeau, J., Liu, S., Sáenz-Moncaleano, C., Sanduvete-Chaves, S., Chacón-Moscoso, S., Becker, B. J., & Tenenbaum, G. (2016). Quiet eye and performance in sport: A meta-analysis. *Journal of Sport & Exercise Psychology, 38*(5), 441–457. https://doi:10.1123/jsep.2015-0123

15. Gallwey, W. T. (2014). *The Inner Game of Tennis.* Pan Macmillan.

16. Causer, J., Vickers, J. N., Snelgrove, R., Arsenault, G., & Harvey, A. (2014). Performing under pressure: Quiet eye training improves surgical knot-tying performance. *Surgery, 156*(5), 1089–1096. https://doi:10.1016/j.surg.2014.05.004

17. Friedman, R. S., Fishbach, A., Förster, J., & Werth, L. (2003). Attentional priming effects on creativity. *Creativity Research Journal, 15*(2–3), 277–286. https://www.researchgate.net/publication/240290779_Attentional_Priming_Effects_on_Creativity

18. Salvi, C., Bricolo, E., Franconeri, S. L., Kounios, J., & Beeman, M. (2015). Sudden insight is associated with shutting out visual inputs. *Psychonomic Bulletin & Review, 22*(6), 1814–1819. https://doi:10.3758/s13423-015-0845-0

19. Chen, X., Cao, L., & Händel, B. F. (2022). Human visual processing during walking: Dissociable pre- and post-stimulus influences. *NeuroImage, 264,* Article 119757. https://doi:10.1016/j.neuroimage.2022.119757

20. Ladouce, S., Donaldson, D. I., Dudchenko, P. A., & Ietswaart, M. (2019). Mobile EEG identifies the re-allocation of attention during real-world activity. *Scientific Reports, 9*(1), Article 15851. https://doi:10.1038/s41598-019-51996-y

21. Chen, X., Cao, L., & Händel, B. F. (2022). Human visual processing during walking: Dissociable pre- and post-stimulus influences. *NeuroImage, 264,* Article 119757. https://doi:10.1016/j.neuroimage.2022.119757

22. Murali, S., & Händel, B. (2022). Motor restrictions impair divergent thinking during walking and during sitting. *Psychological Research, 86*(7), 2144–2157. https://doi:10.1007/s00426-021-01636-w

23. Murali, S., & Händel, B. (2022). Motor restrictions impair divergent thinking during walking and during sitting. *Psychological Research, 86*(7), 2144–2157. https://doi:10.1007/s00426-021-01636-w

24. Davies, M. (2022, February 23). The very capitalist history of the American coffee break. Eater.com. https://www.eater.com/22944907/coffee-break-history-american-work-capitalism

25. Meister, E. (2017). *New York City Coffee: A Caffeinated History*. The History Press.

26. Gonzaga, L. A., Vanderlei, L. C. M., Gomes, R. L., & Valenti, V. E. (2017). Caffeine affects autonomic control of heart rate and blood pressure recovery after aerobic exercise in young adults: A crossover study. *Scientific Reports, 7*(1), Article 14091. https://doi:10.1038/s41598-017-14540-4

27. Bunsawat, K., White, D. W., Kappus, R. M., & Baynard, T. (2015). Caffeine delays autonomic recovery following acute exercise. *European Journal of Preventive Cardiology, 22*(11), 1473–1479. https://doi:10.1177/2047487314554867

28. Lin, Y., Weibel, J., Landolt, H., Santini, F., Slawik, H., Borgwardt, S., Cajochen, C., & Reichert, C. F. (2023). Brain activity during a working memory task after daily caffeine intake and caffeine withdrawal: A randomized double-blind placebo-controlled trial. *Scientific Reports, 13*(1), Article 1002. https://doi:10.1038/s41598-022-26808-5

29. Zabelina, D. L., & Silvia, P. J. (2020). Percolating ideas: The effects of caffeine on creative thinking and problem solving. *Consciousness and Cognition, 79*, Article 102899. https://doi:10.1016/j.concog.2020.102899

Chapter 6: *The Rhythms of the Mind*

1. Kautilya (2016). *The Arthashastra*, L. N. Rangarajan (trans.). Penguin Press. (Original edition published in 1987.)

2. Parkinson's law (1955, November 19). *The Economist*. https://www.economist.com/news/1955/11/19/parkinsons-law

3. Lavie, P. (1992). Ultradian cycles in sleep propensity: Or, Kleitman's BRAC revisited. In D. Lloyd & E. L. Rossi (eds.), *Ultradian rhythms in life processes* (pp. 283–302). Springer. https://doi.org/10.1007/978-1-4471-1969-2_13

4. Fossion, R., Rivera, A. L., Toledo-Roy, J. C., Ellis, J., & Angelova, M. (2017). Multiscale adaptive analysis of circadian rhythms and intra-daily variability: Application to actigraphy time series in acute insomnia subjects. *PLOS ONE, 12*(7), Article e0181762. https://doi.org/10.1371/journal.pone.0181762

5. Blasche, G., Arlinghaus, A., & Crevenna, R. (2022). The impact of rest breaks on subjective fatigue in physicians of the General Hospital of Vienna. *Wiener Klinische Wochenschrift, 134*, 156–161. https://doi.org/10.1007/s00508-021-01949-1

Notes

6. Samuel, I. B. H., Wang, C., Burke, S. E., Kluger, B., & Ding, M. (2019). Compensatory neural responses to cognitive fatigue in young and older adults. *Frontiers in Neural Circuits, 13*, Article 12. https://doi.org/10.3389/fncir.2019.00012

7. Duchniewsk, K., & Kokoszka, A. (2003). The protective mechanisms of the basic rest-activity cycle as an indirect manifestation of this rhythm in waking: Preliminary report. *International Journal of Neuroscience, 113*(2), 153–163. https://doi.org/10.1080/00207450390162001

8. Blasche, G., Khanaqa, T. A. K., & Wagner-Menghin, M. (2023). Mentally demanding work and strain: Effects of study duration on fatigue, vigor, and distress in undergraduate medical students. *Healthcare, 11*(12), Article 1674. https://doi:10.3390/healthcare11121674

9. Salihu, A. T., Hill, K. D., & Jaberzadeh, S. (2022). Neural mechanisms underlying state mental fatigue: A systematic review and activation likelihood estimation meta-analysis. *Reviews in the Neurosciences, 33*(8), 889–917. https://doi:10.1515/revneuro-2022-0023

10. Sun, Y., Lim, J., Dai, Z., Wong, K. F., Taya, F., Chen, Y., Li, J., Thakor, N., & Bezerianos, A. (2017). The effects of a mid-task break on the brain connectome in healthy participants: A resting-state functional MRI study. *NeuroImage, 152*, 19–30. http://doi:10.1016/j.neuroimage.2017.02.084

11. Blasche, G., Zilic, J., & Frischenschlager, O. (2016). Task-related increases in fatigue predict recovery time after academic stress. *Journal of Occupational Health, 58*(1), 89–95. https://doi:10.1539/joh.15-0157-OA

12. Sievertsen, H. H., Gino, F., & Piovesan, M. (2016). Cognitive fatigue influences students' performance on standardized tests. *Proceedings of the National Academy of Sciences, 113*(10), 2621–2624. https://doi:10.1073/pnas.1516947113

13. Schumann, F., Steinborn, M. B., Kürten, J., Cao, L., Händel, B. F., & Huestegge, L. (2022). Restoration of attention by rest in a multitasking world: Theory, methodology, and empirical evidence. *Frontiers in Psychology, 13*, Article 867978. https://doi:10.3389/fpsyg.2022.867978

14. Gao, L., Zhu, L., Hu, L., Hu, H., Wang, S., Bezerianos, A., Li, Y., Li, C., & Sun, Y. (2021). Mid-Task Physical Exercise Keeps Your Mind Vigilant: Evidences From Behavioral Performance and EEG Functional Connectivity. *IEEE transactions on neural systems and rehabilitation engineering : a publication of the IEEE Engineering in Medicine and Biology Society, 29*, 31 -40. https://doi.org/10.1109/TNSRE.2020.3030106

15. Korunka, C., Kubicek, B., Prem, R., & Cvitan, A. (2012). Recovery and detachment between shifts, and fatigue during a twelve-hour shift. *Work, 41*, Supplement 1, 3227–3233. https://doi:10.3233/WOR-2012-0587-3227

Notes

16. Loch, F., Ferrauti, A., Meyer, T., Pfeiffer, M., & Kellmann, M. (2023). Acute effects of mental recovery strategies in simulated air rifle competitions. *Frontiers in Sports and Active Living, 5*, Article 1087995. https://doi:10.3389/fspor.2023.1087995

17. Takahashi, M., Fukuda, H., & Arito H. (1998). Brief naps during post-lunch rest: Effects on alertness, performance, and autonomic balance. *European Journal of Applied Physiology and Occupational Physiology, 78*(2), 93–98. https://doi:10.1007/s004210050392

18. Takahashi, M., Nakata, A., Haratani, T., Ogawa, Y., & Arito, H. (2004). Post-lunch nap as a worksite intervention to promote alertness on the job. *Ergonomics, 47*(9), 1003–1013. https://doi:10.1080/001401304 10001686320

19. Hayashi, M., Chikazawa, Y., & Hori, T. (2004). Short nap versus short rest: Recuperative effects during VDT work. *Ergonomics, 47*(14), 1549–1560. https://doi:10.1080/00140130412331293346

20. Boukhris, O., Trabelsi, K., Ammar, A., Abdessalem, R., Hsouna, H., Glenn, J. M., Bott, N., Driss, T., Souissi, N., Hammouda, O., Garbarino, S., Bragazzi, N. L., & Chtourou, H. (2020). A 90 min daytime nap opportunity is better than 40 min for cognitive and physical performance. *International Journal of Environmental Research and Public Health, 17*(13), Article 4650. https://doi:10.3390/ijerph17134650

21. Dutheil, F., Danini, B., Bagheri, R., Fantini, M. L., Pereira, B., Moustafa, F., Trousselard, M., & Navel, V. (2021). Effects of a short daytime nap on the cognitive performance: A systematic review and meta-analysis. *International Journal of Environmental Research and Public Health, 18*(19), Article 10212. https://doi:10.3390/ijerph181910212

22. Hilditch, C. J., Dorrian, J., Banks, S. (2017). A review of short naps and sleep inertia: Do naps of 30 min or less really avoid sleep inertia and slow-wave sleep? *Sleep Medicine, 32*, 176–190. https://doi:10.1016/j.sleep.2016.12.016

23. Roach, G. D., Matthews, R., Naweed, A., Kontou, T. G., & Sargent, C. (2018). Flat-out napping: The quantity and quality of sleep obtained in a seat during the daytime increase as the angle of recline of the seat increases. *Chronobiology International, 35*(6), 872–883. https://doi:10.1080/07420528.2018.1466801

Chapter 7: *Finding Your Inner Flame*

1. Brown, L. C. (2023). *What Makes You Come Alive: A Spiritual Walk with Howard Thurman*. Broadleaf Books.

Notes

2. Strayer, D. L., Robison, M., & Unsworth, N. (2023, October 23). Effects of goal-setting on sustained attention and attention lapses. *Attention, Perception, & Psychophysics.* https://doi:10.3758/s13414-023-02803-4

3. Baldassarre, G. (2011, August 24-27). What are intrinsic motivations? A biological perspective. In A. Cangelosi, J. Triesch, I. Fasel, K. Rohlfing, F. Nori, P.-Y. Oudeyer, M. Schlesinger, & Y. Nagai (eds.), *Proceedings of the International Conference on Development and Learning and Epigenetic Robotics (ICDL-EpiRob-2011)* (pp. E1–E8). Frankfurt Institute for Advanced Studies. https://doi:10.1109/DEVLRN.2011.6037367

4. Jordet, G., & Hartman, E. (2008). Avoidance motivation and choking under pressure in soccer penalty shootouts. *Journal of Sport & Exercise Psychology, 30*(4), 450–457. https://doi:10.1123/jsep.30.4.450; erratum in Jordet, G., & Hartman, E. (2009). *Journal of Sport & Exercise Psychology, 31*(1), 128–129. https://journals.humankinetics.com/view/journals/jsep/31/1/article-p128.xml

5. Baldassarre, G. (2011, August 24–27). What are intrinsic motivations? A biological perspective. In A. Cangelosi, J. Triesch, I. Fasel, K. Rohlfing, F. Nori, P.-Y. Oudeyer, M. Schlesinger, & Y. Nagai (eds.), *Proceedings of the International Conference on Development and Learning and Epigenetic Robotics (ICDL-EpiRob-2011)* (pp. E1–E8). Frankfurt Institute for Advanced Studies. https://doi:10.1109/DEVLRN.2011.6037367

6. Shannon, C. (1952, March 20). *Creative thinking.* Bell Labs. http://www1.ece.neu.edu/~naderi/Claude%20Shannon.html

7. Kaplan, F., & Oudeyer, P.-Y. (2007). In search of the neural circuits of intrinsic motivation. *Frontiers in Neuroscience, 1*(1), 225–236. https://doi:10.3389/neuro.01.1.1.017.2007

8. Baranès, A. F., Oudeyer, P.-Y., & Gottlieb, J. (2014). The effects of task difficulty, novelty and the size of the search space on intrinsically motivated exploration. *Frontiers in Neuroscience, 8,* Article 317. https://doi.org/10.3389/fnins.2014.00317

9. Howes, A. (2020). *Arts and Minds: How the Royal Society of Arts Changed a Nation.* Princeton University Press.

10. Sloan, A. T., Jones, N. A., & Kelso, J. A. S. (2023). Meaning from movement and stillness: Signatures of coordination dynamics reveal infant agency. *Proceedings of the National Academy of Sciences,* 120(39): Article e2306732120. https://doi.org 10.1073/pnas.2306732120

11. Baranès, A. F., Oudeyer, P.-Y., & Gottlieb, J. (2014). The effects of task difficulty, novelty and the size of the search space on intrinsically motivated exploration. *Frontiers in Neuroscience, 8,* Article 317. https://doi.org/10.3389/fnins.2014.00317

Notes

12. Wilson, R. C., Shenhav, A., Straccia, M., & Cohen, J. D. (2019). The eighty five percent rule for optimal learning. *Nature Communications, 10*(1), Article 4646. https://doi:10.1038/s41467-019-12552-4

13. Watanabe, H., & Naruse, Y. (2022). P300 as a neural indicator for setting levels of goal scores in educational gamification applications from the perspective of intrinsic motivation: An ERP study. *Frontiers in Neuroergonomics, 3*, Article 948080. https://doi:10.3389/fnrgo.2022 .948080

14. Wahab, M., Mead, N., Desmercieres, S., Lardeux, V., Dugast, E., Baumeister, R. F., & Solinas, M. (2023). Cognitive effort increases the intensity of rewards (preprint). https://www.biorxiv.org/content/10 .1101/2023.08.24.554617v1

15. Clay, G., Mlynski, C., Korb, F. M., Goschke, T., & Job, V. (2022). Rewarding cognitive effort increases the intrinsic value of mental labor. *Proceedings of the National Academy of Sciences, 119*(5), Article e2111785119. https://doi:10.1073/pnas.2111785119

Chapter 8: *Finding Your Flow*

1. Conversation with Steve Peters, author of *The Chimp Paradox*, on his podcast *I Am*.

2. Weber, R., Tamborini, R., Westcott-Baker, A., & Kantor, B. (2009). Theorizing flow and media enjoyment as cognitive synchronization of attentional and reward networks. *Communication Theory, 19*(4), 397–422. https://doi.org/10.1111/j.1468-2885.2009.01352.x

3. Chou, Y. (2015). *Actionable Gamification: Beyond Points, Badges, and Leaderboards*. Octalysis Media.

Chapter 9: *Learning at the Pace of Change*

1. McLuhan, M. (1964). *Understanding Media: The Extensions of Man*. McGraw-Hill.

2. Drucker, P. F. (1999). Knowledge-worker productivity: The biggest challenge. *California Management Review, 41*(2), 79–94. https://doi .org/10.2307/41165987

3. Friedman, T. (2017). Thomas Friedman at the 2017 Resnick Aspen Action Forum. Aspen Institute. https://www.youtube.com/watch?v =_5vGQQYuJBo

4. McNab, F., Zeidman, P., Rutledge, R. B., Smittenaar, P., Brown, H. R., Adams, R. A., & Dolan, R. J. (2015). Age-related changes in working memory and the ability to ignore distraction. *Proceedings of the*

National Academy of Sciences, 112(20), 6515–6518. https://doi:10.1073/pnas.1504162112

5. Unsworth, N., & Robison, M. K. (2017). A locus coeruleus-norepinephrine account of individual differences in working memory capacity and attention control. *Psychonomic Bulletin & Review, 24,* 1282–1311. https://doi.org/10.3758/s13423-016-1220-5

6. Wamsley, E. J., & Summer, T. (2020). Spontaneous entry into an "offline" state during wakefulness: A mechanism of memory consolidation? *Journal of Cognitive Neuroscience, 32*(9), 1714–1734. https://doi.org/10.1162/jocn_a_01587

7. Schmidt-Kassow, M., Zink, N., Mock, J., Thiel, C., Vogt, L., Abel, C., & Kaiser, J. (2014). Treadmill walking during vocabulary encoding improves verbal long-term memory. *Behavioral and Brain Functions, 10,* Article 24. https://doi:10.1186/1744-9081-10-24

8. Judde, S., & Rickard, N. (2010). The effect of post-learning presentation of music on long-term word-list retention. *Neurobiology of Learning and Memory, 94*(1), 13–20. https://doi:10.1016/j.nlm.2010.03.002

9. Nielson, K. A., & Powless, M. (2007). Positive and negative sources of emotional arousal enhance long-term word-list retention when induced as long as 30 min after learning. *Neurobiology of Learning and Memory, 88*(1), 40–47. https://doi:10.1016/j.nlm.2007.03.005

10. Nielson, K. A., Yee, D., & Erickson, K. I. (2005). Memory enhancement by a semantically unrelated emotional arousal source induced after learning. *Neurobiology of Learning and Memory, 84*(1), 49–56. https://doi:10.1016/j.nlm.2005.04.001

11. Nielson, K. A., & Arentsen, T. J. (2012). Memory modulation in the classroom: Selective enhancement of college examination performance by arousal induced after lecture. *Neurobiology of Learning and Memory, 98*(1), 12–16. https://doi:10.1016/j.nlm.2012.04.002

12. Dembitskaya, Y., Piette, C., Perez, S., Berry, H., Magistretti, P. J., & Venance, L. (2022). Lactate supply overtakes glucose when neural computational and cognitive loads scale up. *Proceedings of the National Academy of Sciences, 119*(47), Article e2212004119. https://doi:10.1073/pnas.2212004119

13. Dennis, A., Thomas, A. G., Rawlings, N. B., Near, J., Nichols, T. E., Clare, S., Johansen-Berg, H., & Stagg, C. J. (2015). An ultra-high field magnetic resonance spectroscopy study of post exercise lactate, glutamate and glutamine change in the human brain. *Frontiers in Physiology, 6,* Article 351. https://doi:10.3389/fphys.2015.00351

14. van Dongen, E. V., Kersten, I. H. P., Wagner, I. C., Morris, R. G. M., & Fernández, G. (2016). Physical exercise performed four hours after

learning improves memory retention and increases hippocampal pattern similarity during retrieval. *Current Biology, 26*(13), 1722–1727. https://doi:10.1016/j.cub.2016.04.071

15. Winter, B., Breitenstein, C., Mooren, F. C., Voelker, K., Fobker, M., Lechtermann, A., Krueger, K., Fromme, A., Korsukewitz, C., Floel, A., & Knecht, S. (2007). High impact running improves learning. *Neurobiology of Learning and Memory, 87*(4), 597–609. https://doi:10.1016/j.nlm.2006.11.003

16. Schmidt-Kassow, M., Deusser, M., Thiel, C., Otterbein, S., Montag, C., Reuter, M., Banzer, W., & Kaiser, J. (2013). Physical exercise during encoding improves vocabulary learning in young female adults: A neuroendocrinological study. *PLOS ONE, 8*(5), Article e64172. https://doi:10.1371/journal.pone.0064172

17. Kim, Y., Sidtis, D. V. L., & Sidtis, J. J. (2021). Emotional nuance enhances verbatim retention of written materials. *Frontiers in Psychology, 12*, Article 519729. https://doi:10.3389/fpsyg.2021.519729

Chapter 10: *The Creative Mind*

1. Ogilvy, D. (1983). *Ogilvy on Advertising*. Crown.
2. Wallas, G. (1926). *The Art of Thought*. Harcourt, Brace and Company.
3. Ishiguro, K. (2014, December 6). How I wrote *The Remains of the Day* in four weeks. *The Guardian*. https://www.theguardian.com/books/2014/dec/06/kazuo-ishiguro-the-remains-of-the-day-guardian-book-club
4. Ibid.
5. Nijstad, B. A., De Dreu, C. K. W., Rietzschel, E. F., & Baas, M. (2010). The dual pathway to creativity model: Creative ideation as a function of flexibility and persistence. *European Review of Social Psychology, 21*, 34–77. https://doi:10.1080/10463281003765323
6. University of Minnesota. (2007, April 25). Ceiling height can affect how a person thinks, feels and acts. *ScienceDaily*. sciencedaily.com/releases/2007/04/070424155539.htm
7. Martindale, C. (1981). *Cognition and Consciousness*. Dorsey Press.
8. Memmert, D. (2007). Can creativity be improved by an attention-broadening training program? An exploratory study focusing on team sport. *Creativity Research Journal, 19*(2–3), 281–291. https://doi.org/10.1080/10400410701397420
9. The Video Arts lecture, no longer available online, was given on January 23, 1991, at Grosvenor House Hotel, London, United King-

dom. But see Cleese, J. [@JohnCleese]. (2023, July 25). *More on the creative process and Alfred Hitchcock . . .* [tweet]. Twitter. https://twitter .com/i/status/1683920371429552129

10. Gertner, J. (2012, February 25). True innovation. *The New York Times.* https://www.nytimes.com/2012/02/26/opinion/sunday/innovation -and-the-bell-labs-miracle.html

11. Yang, Z., Hung, I. W. (2021). Creative thinking facilitates perspective taking. *Journal of Personality and Social Psychology, 120*(2), 278–299. https://doi:10.1037/pspa0000259

12. Waldrop, M. (2017, February 3). Inside Einstein's love affair with "Lina" — his cherished violin. *National Geographic.* https://www .nationalgeographic.com/adventure/article/einstein-genius-vio lin-music-physics-science

Chapter 11: *Creaking Under the Information Load*

1. McLuhan, M. (1964). *Understanding Media: The Extensions of Man.* McGraw-Hill.

2. Dixon, S. J. (2023, August 23). *Countries with highest number of emails sent 2023.* Statista. https://www.statista.com/statistics/1270459 /daily-emails-sent-by-country/

3. Sweller, J., van Merriënboer, J. J. G., & Paas, F. (2019). Cognitive architecture and instructional design: 20 years later. *Educational Psychology Review, 31*, 261–292. https://doi.org/10.1007/s10648-019 -09465-5

4. Sinclair, A. H., Wang, Y. C., & Adcock, R. A. (2023). Instructed motivational states bias reinforcement learning and memory formation. *Proceedings of the National Academy of Sciences, 120*(31), Article e2304881120. https://doi:10.1073/pnas.2304881120

5. Daft, R. L., & Lengel, R. H. (1986). Organizational information requirements, media richness and structural design. *Management Science, 32*(5), 554–571. http://www.jstor.org/stable/2631846

6. de Gee, J. W., Colizoli, O., Kloosterman, N. A., Knapen, T., Nieuwenhuis, S., & Donner, T. H. (2017). Dynamic modulation of decision biases by brainstem arousal systems. *eLife, 6*, Article e23232. https:// doi:10.7554/eLife.23232

7. Murphy, P. R., Boonstra, E., & Nieuwenhuis, S. (2016). Global gain modulation generates time-dependent urgency during perceptual choice in humans. *Nature Communications, 7*, Article 13526. https:// doi:10.1038/ncomms13526

Notes

8. Shapiro, J. (2003, January 24). *Bombing hearing highlights military's amphetamine policy* [Radio broadcast]. NPR. https://www.npr.org/2003/01/24/935462/bombing-hearing-highlights-militarys-amphetamine-policy

9. Wills, M. (2022, September 17). The RAF on speed: High-flying or flying high? JSTOR Daily. https://daily.jstor.org/the-raf-on-speed-high-flying-or-flying-high/

10. Cooke, R. (2016, September 25). High Hitler: How Nazi drug abuse steered the course of history. *The Guardian.* https://www.theguardian.com/books/2016/sep/25/blitzed-norman-ohler-adolf-hitler-nazi-drug-abuse-interview

11. De Couck, M., Caers, R., Musch, L.,Fliegauf, J., Giangreco, A., & Gidron, Y. (2019). How breathing can help you make better decisions: Two studies on the effects of breathing patterns on heart rate variability and decision-making in business cases International journal of psychophysiology : official journal of the International Organization of Psychophysiology, 139, 1–9. https://doi.org/10.1016/j.ijpsycho.2019.02.011

12. Baer, T., & Schnall, S. (2021). Quantifying the cost of decision fatigue: Suboptimal risk decisions in finance. *Royal Society Open Science, 8,* Article 201059. https://doi.org/10.1098/rsos.201059

13. Allan, J. L., Johnston, D. W., Powell, D. J. H., Farquharson, B., Jones, M. C., Leckie, G., & Johnston, M. (2019). Clinical decisions and time since rest break: An analysis of decision fatigue in nurses. *Health Psychology: Official Journal of the Division of Health Psychology, American Psychological Association, 38*(4), 318–324. https://doi.org/10.1037/hea0000725

14. Bae, S., & Masaki, H. (2019). Effects of acute aerobic exercise on cognitive flexibility required during task-switching paradigm. *Frontiers in Human Neuroscience, 13,* Article 260. https://doi:10.3389/fnhum.2019.00260

15. Temple University. (2001, July 30). Let's face it, man is not made to communicate electronically. *ScienceDaily.* sciencedaily.com/releases/2001/07/010730081336.htm

16. Zivan, M., Vaknin, S., Peleg, N., Ackerman, R., & Horowitz-Kraus, T. (2023). Higher theta-beta ratio during screen-based vs. printed paper is related to lower attention in children: An EEG study. *PLOS ONE, 18*(5), Article e0283863. https://doi:10.1371/journal.pone.0283863

17. Harris, D. J., Buckingham, G., Wilson, M. R., & Vine, S. J. (2019). Virtually the same? How impaired sensory information in virtual reality may disrupt vision for action. *Experimental Brain Research, 237*(11), 2761–2766. https://doi:10.1007/s00221-019-05642-8

18. Juliano, J. M., Schweighofer, N., & Liew, S. (2022). Increased cognitive load in immersive virtual reality during visuomotor adaptation is associated with decreased long-term retention and context transfer. *Journal of NeuroEngineering and Rehabilitation, 19*(1), Article 106. https://doi:10.1186/s12984-022-01084-6

Chapter 12: *Too Little Time*

1. Beard, G. M. (1881). *American Nervousness: Its Causes and Consequences.* G. P. Putnam's Sons.
2. Pappalardo, J. (2011, October 27). New transatlantic cable built to shave 5 milliseconds off stock trades. *Popular Mechanics.* https://www.popularmechanics.com/technology/infrastructure/a7274/a-transatlantic-cable-to-shave-5-milliseconds-off-stock-trades/
3. Johnson, N., Zhao, G., Hunsader, E., Qi, H., Johnson, N., Meng, J., & Tivnan, B. (2013). Abrupt rise of new machine ecology beyond human response time. *Scientific Reports, 3,* Article 2627. https://doi:10.1038/srep02627
4. Graeber, T., Roth, C., & Zimmermann, F. (2022). DP17683 Stories, Statistics, and Memory. Discussion Paper No. 17683. CEPR Press. https://cepr.org/publications/dp17683
5. Hassan, R. (2008). *The Information Society: Cyber Dreams and Digital Nightmares.* Wiley.
6. Schumann, F., Steinborn, M. B., Kürten, J., Cao, L., Händel, B. F., & Huestegge, L. (2022). Restoration of attention by rest in a multitasking world: Theory, methodology, and empirical evidence. *Frontiers in Psychology, 13,* Article 867978. https://doi:10.3389/fpsyg.2022.867978
7. Ibid.
8. Wiener, E. L., Curry, R. E., & Faustina, M. L. (1984). Vigilance and task load: In search of the inverted U. *Human Factors, 26*(2), 215–222. https://doi.org/10.1177/001872088402600208
9. Chan, T. F. (2018, May 1). China is monitoring employees' brain waves and emotions—and the technology boosted one company's profits by $315 million. *Business Insider.* https://www.businessinsider.com/china-emotional-surveillance-technology-2018-4?r=US&IR=T
10. Robison, M. K., Unsworth, N., & Brewer, G. A. (2021). Examining the effects of goal-setting, feedback, and incentives on sustained attention. *Journal of Experimental Psychology: Human Perception and Performance, 47*(6), 869–891. https://doi:10.1037/xhp0000926
11. Pop, V. L., Stearman, E. J., Kazi, S., & Durso, F. T. (2012). Using engagement to negate vigilance decrements in the NextGen environment.

International Journal of Human–Computer Interaction, 28(2), 99–106. https://doi:10.1080/10447318.2012.634759

12. Robison, M. K., Unsworth, N., & Brewer, G. A. (2021). Examining the effects of goal-setting, feedback, and incentives on sustained attention. *Journal of Experimental Psychology: Human Perception and Performance, 47*(6), 869–891. https://doi:10.1037/xhp0000926

13. Novakovic-Agopian, T., Kornblith, E., Abrams, G., Burciaga-Rosales, J., Loya, F., D'Esposito, M., Chen, A. J. W. (2018). Training in goal-oriented attention self-regulation improves executive functioning in veterans with chronic traumatic brain injury. *J Neurotrauma, 35*(23), 2784–2795. doi: 10.1089/neu.2017.5529

14. O'Connell, R. G., Bellgrove, M. A., Dockree, P. M., Lau, A., Fitzgerald, M., Robertson, I. H. (2008). Self-alert training: Volitional modulation of autonomic arousal improves sustained attention. *Neuropsychologia, 46*(5), 1379–90. https://doi.org/10.1016/j.neuropsychologia.2007.12.018

Chapter 13: *An Uncertainty Paradox*

1. Pascal, B. (1814). In Armand-Prosper Faugère (ed.), *Pensées, fragments et lettres de Blaise Pascal* (Vol. 2, p. 99). Andrieux.

2. Hilbert, M., & Darmon, D. (2020). How complexity and uncertainty grew with algorithmic trading. *Entropy, 22*(5), Article 499. https://doi .org/10.3390/e22050499

3. Evans, D., Pottier, C., Fletcher, R., Hensley, S., Tapley, I., Milne, A., & Barbetti, M. (2007). A comprehensive archaeological map of the world's largest preindustrial settlement complex at Angkor, Cambodia. *Proceedings of the National Academy of Sciences, 104*(36), 14277–14282. https://doi:10.1073/pnas.0702525104

4. Izrailevsky, Y., & Tseitlin, A. (2011, July 19). The Netflix simian army. Netflix technology blog. https://netflixtechblog.com/the-net flix-simian-army-16e57fbab116

5. Berlyne, D. E. (1960). *Conflict, Arousal and Curiosity*. McGraw-Hill.

6. Garber, G. (2012, June 7). Why Nadal needs to break his habits. ESPN. https://www.espn.com/tennis/french12/story/_/id/8019942/french -open-why-rafael-nadal-needs-break-habits

Conclusion: *From Marching Soldier to Spinning Dancer*

1. McLuhan, M. (1964). *Understanding Media: The Extensions of Man.* McGraw-Hill.

INDEX

Index

Index

Index

Index

Index

Index

Index

DR. MITHU STORONI is a University of Cambridge-trained physician, neuroscience researcher and ophthalmic surgeon. She advises multinational corporations on mental performance and stress management.